To Dr. W. M

With kindest regards and best wishes

R. S. Young

December, 1979

Cobalt in
Biology and Biochemistry

Cobalt in Biology and Biochemistry

ROLAND S. YOUNG

Consulting Chemical Engineer, Victoria, B.C., Canada

1979

ACADEMIC PRESS

London · New York · San Francisco

A Subsidiary of Harcourt Brace Jovanovich, Publishers

ACADEMIC PRESS INC. (LONDON) LTD.
24/28 Oval Road
London NW1

United States Edition published by
ACADEMIC PRESS INC.
111 Fifth Avenue
New York, New York 10003

British Library Cataloguing in Publication Data

Young, Roland Stansfield
  Cobalt in biology and biochemistry.
  1. Cobalt in the body
  2. Plants, Effect of cobalt on
  574.1'9214   QP535.C6   79-50304
  ISBN 0-12-772750-7

Typeset in Great Britain by Eta Services (Typesetters) Ltd.
and printed in Great Britain.

Preface

The essential role of cobalt in ruminant nutrition, its function as a key element in vitamin B_{12}, and the extensive evidence on the beneficial effects of small quantities of this metal on growth and development in many plants, animals, and microorganisms, place cobalt in a unique position among the rarer heavy metals which are found in biological systems.

Information on cobalt in biology and biochemistry is scattered among publications devoted to soils, plant physiology, animal science, microbiology, foods and nutrition, biochemistry, veterinary science, physiology, waters, toxicology and industrial hygiene, environmental studies, enzymes, geochemistry, and many other subjects.

By assembling in this little volume a summary of this information it is our hope that investigators may benefit from an insight into the techniques employed, and the data obtained, by others in diverse fields of science who are likewise interested in the occurrence and role of cobalt in biology and biochemistry.

April 1979

R. S. YOUNG

Contents

1 Introduction

The importance of cobalt in biology and biochemistry was first realized over forty years ago when the cause of puzzling disorders of cattle and sheep, in various parts of the world, was finally traced to a deficiency of this element in their diet. Studies gradually showed that the action of cobalt was associated with the ruminant stomach, and that rumen microorganisms played a role in cobalt metabolism. After it was found that vitamin B_{12} was a cobalt-containing substance, many investigators confirmed that this vitamin is an important intermediary in cobalt metabolism in ruminants; most workers now believe that cobalt deficiency in ruminants is essentially a deficiency of vitamin B_{12}.

Cobalt constitutes 4.35% by weight of vitamin B_{12}, and to date is the only metal found in a vitamin. About 1 μg of vitamin B_{12} per day is required in human nutrition. Although in most countries pernicious anemia has been virtually eliminated, or at least controlled, by vitamin B_{12} therapy, the precise nature of the mechanism by which this has been brought about is still not known.

In spite of the amount of work carried out on the effect of cobalt on plants, it is still not certain that this element can be classed as essential for plant nutrition. The majority of investigators have found that the addition of cobalt compounds to plants results in varying degrees of growth stimulation, but it is not yet known whether or not plants are able to grow in the absence of traces of cobalt.

Over many years, a number of papers have discussed the effect of cobalt on algae, bacteria, fungi, and other microorganisms, and on various enzyme activities. As with plants, the addition of a small quantity of cobalt frequently stimulates growth or activity, whereas a higher concentration can exert an inhibiting effect; whether or not the element is essential, and, if so, at what level, is still a subject of controversy.

In succeeding chapters, the occurrence and biochemical role of cobalt in such biological media as soils, plants, animals, humans, microorganisms, waters and wastewaters, fertilizers, and others, will be reviewed. Before proceeding, however, it will be useful to discuss briefly the original source of cobalt for all living matter—that occurring in rocks and meteorites.

The igneous rocks of the earth's surface have been calculated by different authorities to contain about 0·001,[1] 0·002,[2] and 0·004 %[3] cobalt. The element is usually assigned about thirty-third position in order of abundance in the outer crust of the earth. In general, cobalt may be found in rocks to the extent of about 0·2–250 mg kg^{-1}.[4] A number of values for cobalt concentrations in various rocks may be found in a comprehensive handbook of geochemistry.[5]

Magmatic differentiation, which produces basic or ultrabasic rocks, appears to concentrate cobalt along with nickel, chromium, and some other elements. Usually the cobalt content of rocks which crystallize first from a molten magma is higher than that of rocks which solidify last.[6–10] Cobalt enters predominantly into ferromagnesian silicates on crystallization of the original magma.[11] The element is generally low in granites, typically 0·5–5 mg kg^{-1}.[12] The valuable cobalt content of the vast nickel-containing laterites and limonites is derived from the weathering of ultramafic rocks, chiefly peridotites and serpentines, which contain about 200 mg cobalt kg.$^{-1}$ A number of investigators have reported a low level of cobalt in limestones, varying from 0·2–12·5 mg kg^{-1};[13] a Brazilian team has recently reported a content ranging from trace levels to 46·9 mg kg^{-1}, with an average of 4·2 mg kg^{-1}.[14]

Cobalt is found in meteorites as either the metal or the oxide. The cobalt content of the silicate phase of some stony meteorites is roughly comparable to that in many rocks; figures have been published showing about 40 mg kg^{-1},[15] 200 mg kg^{-1},[16] and 700 mg kg^{-1}.[17] The metal phase of stony meteorites, and iron meteorites, of course, have much higher cobalt contents, ranging from about 0·37–1·63 %.

Apart from the cobalt derived originally from rocks and meteorites, traces of this element may be distributed to soils, plants, animals, humans, microorganisms, waters, etc. from man-made substances containing cobalt. This element is one of the principal pigments imparting a blue color to glass, pottery, porcelain, and china; cobalt is also the best element for promoting the adhesion of enamel to steel, and is used extensively as a ground coat in kitchen enamelware. Electrolytic nickel contains a small quantity of cobalt, and traces of the latter may consequently be encountered in food-processing operations, where nickel, Monel, or stainless steel are employed either in food manufacture or in kitchens. Cobalt compounds are frequently used as driers for paints, varnishes, and lacquers; it has been suggested that minute quantities of the metal may be derived from driers for lacquers in cans, and for paints used on the inner surfaces of fish boxes.[18] Other sources of trace quantities of cobalt are contaminants and dusts from industrial products such as high-speed steel, cemented carbides, magnets, high-temperature alloys, nickel electroplating, and catalysts.

Full details of cobalt mineral deposits, extractive metallurgy, physical and chemical properties of cobalt, preparation and properties of its compounds, and uses of cobalt are covered in monographs[13,19,20] and texts.[21,22]

References, numbered in chronological order, are listed separately for each chapter at the end of the latter. For a subject which includes many diverse facets, this practice provides a desirable proximity for text and references.

References

1. Clarke, F. W. (1916). "The Data of Geochemistry". U.S. Geol. Sur. Bull. 616.
2. Mason, B. (1952). "Principles of Geochemistry". John Wiley, New York.
3. Lundegardh, P. H. (1951). *Sveriges Geol. Undersokn, Arsbok Ser. C Avhandl och Uppsal* No. 513, 43, No. 11; (1951). *C.A.** **45**, 88.
4. Young, R. S. (1957). *Geochim. Cosmochim. Acta* **13**, 28–41.
5. Wedepohl, K. H. (1969–72). "Handbook of Geochemistry", Vols 1–4. Springer-Verlag, Berlin.
6. Wager, L. R. and Mitchell, R. L. (1943). *Mining Mag.* **26**, 283–96.
7. Lundegardh, P. H. (1945). *Nature, Lond.* **155**, 753.
8. Nockolds, S. R. and Mitchell, R. L. (1948). *Trans. R. Soc., Edinburgh* **61**, 533–75.
9. Wager, L. R. and Mitchell, R. L. (1951). *Geochim. Cosmochim. Acta* **1**, 129–208.
10. Wager, L. R. and Mitchell, R. L. (1953). *Geochim. Cosmochim. Acta* **3**, 217–23.
11. Vogt, J. H. L. (1926). *Econ. Geol.* **21**, 207–33, 309–32, 469–97.
12. Sandell, E. B. and Goldich, S. S. (1943). *J. Geol.* **51**, 99–115, 167–89.
13. Young, R. S. (1960). "Cobalt", A.C.S. Monograph 149. Reinhold, New York.
14. Valadares, J. M. A. S., Bataglia, O. C., and Furlani, P. R. (1974). *Brazantia* **33**, 147–52; (1975). *C.A.* **83**, 8239.
15. Beck, C. W. and La Paz, L. (1951). *Am. Miner.* **36**, 45–59.
16. Brown, H. and Patterson, C. (1947). *J. Geol.* **55**, 405–11, 508–10.
17. Wahl, W. (1950). *Mineral. Mag.* **24**, 416–26.
18. Monier-Williams, G. W. (1949). "Trace Elements in Foods". Chapman and Hall, London.
19. Perel'man, F. M., Zvorykin, A. Y., and Gudima, N. V. (1949). "Kobal't". Akad. Nauk S.S.S.R., Moscow.
20. Andrews, R. W. (1962). "Cobalt". H.M. Stationery Office, London.
21. Mellor, J. W. (1935). "A Comprehensive Treatise on Inorganic and Theoretical Chemistry", Vol. XIV. Longmans, Green, London.
22. Sidgwick, N. V. (1950). "The Chemical Elements and Their Compounds", Vol. II. Clarendon Press, Oxford.

* Chemical Abstracts.

2 Cobalt in Soils

Cobalt is found is most soils in the range 0.1–50 mg kg^{-1}, or parts 10^{-6}, and this small quantity failed to arouse interest among soil investigators, with the exception of Bertrand[1] and McHargue,[2] until about 1934, when various ruminant disorders were linked to a deficiency of cobalt. Since then, soils in a large number of localities throughout the world have been found to contain too little cobalt to maintain the good health of the cattle and sheep subsisting on the pasture or crops of these areas. The addition of small quantities of cobalt to either feed, pasture, water, salt-lick, fertilizer, or limestone is now an established practice in many regions.

In Table 1 are tabulated many data on total and available cobalt in soils reported in the literature in recent years. There is a wide variation in cobalt content, as would be expected in soils derived from different parent materials, and subjected to varying physical, chemical, and biological activities. Total cobalt may range from about 0.3 mg kg^{-1} of soil in markedly cobalt-deficient areas to 1000 mg kg^{-1} soil over mineralized areas; most soils fall within a range of 2–40 mg Co kg^{-1} of soil. Available cobalt varies from about 0.01–6.8 mg kg^{-1} of soil, but is usually in a range of 0.1–2 mg kg^{-1}.

Total cobalt is the amount found in the sample after complete decomposition by either acid treatment or fusion. Available cobalt is that portion which is extracted from the soil sample by treatment with dilute acids or salt solutions. Though it has been reported that the cobalt content of plants depended more on the total than on the available cobalt in soils,[49] most workers believe that available cobalt can be more readily correlated with a deficiency of this element.[94,116,125,134,173,174] Available cobalt as a percentage of total cobalt shows a very wide range, from 1–93%, but is usually between 3 and 20%.

Many extractants have been recommended for the determination of available cobalt in soils. Among those reported have been 1 M hydrochloric acid,[173,175] 0.01 M hydrochloric acid,[176] 10% by vol. hydrochloric acid,[78] 1 M potassium nitrate at pH 3,[177] 0.1 M calcium chloride,[176] 2% citric acid,[173] 0.2 M oxalic acid,[173] 0.01 M EDTA,[178] 0.02 M EDTA,[15] 1 M nitric acid,[179,180] 0.1 M nitric acid,[181] ammonium acetate.[28,182] The most popular extractant is 2.5% by vol. acetic acid.[12,18,174,183–185]

Table 1 Cobalt Content of Soils

Soil	Cobalt (mg kg^{-1})		
	Total	Available	Ref.
Australia			
Unhealthy for sheep	0·1–1·5		3
Healthy for sheep	0·2–32		3
Kraznozems and lateritic kraznozems	4–82		4
Brown earths, black earths and prairie soils	11–94		4
Brazil			
Deficient	<2·5		5
Moderately deficient	2·5–5		5
Deficient	<2·5	<0·1	6
Britain			
Moorland, unhealthy	3·9		7,8
Lowland, healthy	16·7		7,8
Devon, unhealthy	4		7,8
Devon, healthy	19·6		7,8
Dartmoor, "pining"	0·20		9
Dartmoor, healthy	0·45		9
Cornwall, "pining"		0·033	8
Cornwall, healthy		1·211	8
Canada			
Various	0·8–18·2		10
Nova Scotia	3·6–21		11
Quebec	1·1–21·6	0·3–0·83	12
Ontario	4·4		13
Cuba			
Laterites	427–466		14
Denmark			
Surface horizon		0·95	15
Clay		0·39	15
Sandy		0·078	15
Dominican Republic			
Laterite	40		16
Egypt			
Nile sediment	100–130		17

Table 1 Cobalt Content of Soils (cont'd)

Soil	Cobalt (mg kg^{-1})		Ref.
	Total	Available	
France			
Derived from granite	2·2–3·8		1
Fertile garden	37		1
Granitic rock area		0·28	18
Shale		0·56	18
Germany			
Schwarzwald, surface		0·107	16
Black Forest, sick farm	0·05		19,20
Black Forest, healthy farm	0·14		19,20
Braunerde	4–8·5		21
Marsh	8–11·5		22
Podzols	0·5–1·8		22
Low Moor	0·9–3·1		22
Haiti			
Laterite	40		16
Holland			
Cobalt-deficient	0·3		23
Sandy	0·3		24
India			
Desert	5		25
Semi-arid	8·5–18·5		25
Gujarat		0·25	26
Gujarat	12–48	0·10–0·60	27
Konkan	38–68	0·056–0·46	28
Gujarat	4–78	0·1–2·1	29
Uttar Pradesh, alkali	4·6–29·1	0·06–0·56	30
Uttar Pradesh, normal	5·6–28·6	0·09–0·48	30
Italy			
Various	4·59–20		31
Various	0·13–15·6		32
Alpine slopes	0·05–0·6		33
Toscana	8·17	0·41	34
Pavia	11·5–34		35
Japan			
Soil	35		16
Kenya			
East-African		0·58	36

Table 1 (cont'd)

Soil	Cobalt (mg kg^{-1})		Ref.
	Total	Available	
Malawi			
Red and black	15–27		16
Malaya			
Various	0–2		16
Mongolia			
Chestnut	0·8–2·3		37
Soils	3·5–38	0·19–2·84	38
Soils	3·5–38	0·25–2	39
New Zealand			
"Bush-sick"	0·05–0·23		40
Healthy	0·33–0·94		40
Unhealthy	0·4–2·5		41,42
Normal	2·8–25·4		41,42
Morton Mains, healthy	2·8–8·3		41,42
Morton Mains, unhealthy	3·3–4·8		41,42
Various	0·4–385		16
Nigeria			
Soils	6·3–13		16
Norway			
Soils	0·05–5		16
Pakistan			
West-Pakistan	12		43
Reclaimed land	2		44
Unreclaimed land	5–8		44
Poland			
Peat	1		45
Sandy	tr.–5		45
Loam	10–20		45
Grass meadow	0·37–19·5		46
Brown clay	8·2–15·5		47,48
Loam	14·4–28·7		47,48
Sandy	7		47,48
Peat	1·1–1·9	0·03–0·07	49
Black earths	5·0–6·5	0·16–0·19	49
Podzolic	2·9–6·9	0·08–0·26	49
Brown	3·6–6·1	0·07–0·24	49

Table 1 Cobalt Content of Soils (cont'd)

Soil	Cobalt (mg kg^{-1})		Ref.
	Total	Available	
Poland (cont'd)			
Silesian sandy	9	0·22	50
Black earth	5·41	0·15	51
Brown	5·17	0·14	51
Cultivated podzolic	2·91	0·09	51
Piaseczno District	1·03–8·6	0·01–0·675	52
Romania			
Soil	4·1–11·4		53
Peat	0–4·5		53
Chernozems	1·22–10·66		54
Various	0·44–14·77	0·12–3·36	54
Soil	5		55
Danube delta	2·2–15	0·10–2·70	56
Spain			
Various	0·1–20		57
Sweden			
Various	0·23–1·84		16
Taiwan			
Soil	0·72		58
Uganda			
Soil		0·02–0·78	59
United States of America			
Kentucky, virgin	1·5		2
Kentucky, various		0·2	60
Missouri	4·2–37		61
New Jersey	0·2–30·8		62
Grey–brown podzolic, ex loess	1·1–1·7		63
Grey–brown podzolic, ex shale	0·1–0·4		63
Chernozem	0·1–0·2		63
Prairie	0·1–2·4		63
Chernozem prairie	0–1·9		63
Lateritic, ex limestone	0·2–0·9		63
Lateritic, ex heavy clay	0·6–1·0		63
Rendzina	0·9–1·4		63
Ground-water podzols	5		64
Regosols, red–yellow podzols, humic-gleys, red–brown lateritic	10–295		64
Granitic drift deposits	5		65

Table 1 (cont'd)

Soil	Cobalt (mg kg^{-1})		Ref.
	Total	Available	
U.S.S.R.			
Latvia, ploughed layer	0·4–4		66
Light podzolic, deficient	0·3–1·5		67
Sandy light, deficient	0·4–1·5		68
Various	1·7–5·4		69
Peat podzol	1		70
Above ore deposit	1000		70
Urals, chestnut brown	20–430		70
Clays	11		71
Ukraine, chernozem	2·3		72
Ukraine, acid podzolic and forest	0·25		72
Clay	2·5–4·0		73
Carbonaceous, low in humus	0–5		74
Acid sandy	0·1–1·8		74
Various	0·12–6		75
Chkalov region, chernozem	0·04–19		76
Chkalov region, chernozem	0·45–4·5		77
Lithuania, surface soil	0·6–3·2		78
U.S.S.R.	0·4–360		79
Tedzhen Delta, sandy	10–30		80
Opol'ya, dark grey, meadow, meadow gley, and loess-like	5		81
Opol'ya, chernozem	8		81
Urals, swamp-forest	<10		82
Urals, forest	10		82
Urals, forest steppe	10–30		82
Kokshaga	14–45		83
Klyazma	tr.–34		83
Pripyat	3–5		83
Karelia, deficient	1		84
Azerbaidzhan, vineyards	0·6–28·4		85
Stavropol	0·51–8·7		86
Chelyabinsk region, podzol-carbonate, sod-podzolic, and degradated chernozem	2·1–3·7		87
Polessie	0·11–6		88
Grey, irrigated	6–12		89
Grey, non-irrigated	2·9–12·4		89
Meadow, irrigated	5·4–11·7		89
Swamp-meadow, irrigated	10–10·7		89
Russia, average	10		90
Low goiter morbidity	1·83		90
High goiter morbidity	1·08		90

Table 1   Cobalt Content of Soils (cont'd)

	Cobalt (mg kg^{-1})		
Soil	Total	Available	Ref.
U.S.S.R. (cont'd)			
Ruminant deficiency	<1·5		90
Brown-meadow podzolized	0·807–0·905		91
Humus-gley alluvial	2·494		91
Top-soil horizon, max.		0·714–1·894	91
Brown and dark grey, unirrigated	16·7	1·65	92
Grey, irrigated	11·2	1·65	92
Meadow-grey, irrigated	12·4	1·11	92
Light grey, irrigated	5	0·6	92
Solonchaks in old cotton soil	11	0·85	92
Soil-forming loess	11	—	92
Sierozems, unirrigated	4·4–12·3		93
Sierozems, irrigated	6·2–12·4		93
Meadow, irrigated	3·6–11·7		93
Ukraine, various	tr.–5·8		94
Meadow-like black		0·64–1·30	95
Normal black		1·8–1·96	95
Podzolized black		0·46–0·68	95
Low-humus black		0·64–1·08	95
Turf middle-podzolized		0·8–1·0	95
Grey forest podzol		0·88–1·1	95
Mogilev region, loam	2·29–5·64	0·51–0·99	96
Sandy loam	1·14–4·36	0·11–0·31	96
Sandy	0·337–0·89	0·02–0·20	96
Krasnodar region, chernozem		1·5–2·8	97
Kalinin region, soddy-podzolic	2–10		98
Central Vetluga area	3·75–18·5	0·1–2·0	99
Ukraine		2·5–4·5	100
Chelyabinsk region	1·67–17·56		101
Leningrad region, deficient	0·08–0·96		102
Salyansk region		0·008–1·6	103
Armenian S.S.R.		3	104
Karelian S.S.R.	1·2–3·8		105
Sartavala district	1·2–17·5		106
Shemakhin district		tr.–0·3	107
Uzbek, sandy desert	10		108
Komsomol state farm	61·3		109
Salyany area		0·008–1·66	110
Semi-desert and gypsum desert	8·6	1·2	111
Azerbaidzhan	1·5		112
Kryazh, state farm	30		113
Brown, semi-desert		2·0–2·5	114
Chestnut		2·4–4·5	114

Table 1   (cont'd)

Soil	Cobalt (mg kg^{-1})		Ref.
	Total	Available	
U.S.S.R. (cont'd)			
Chernozems		high	114
Brown mountain-forest		1·5–3·5	114
Mountain-meadow		2·0–6·1	114
Azerbaidzhan, pasture area	2	0·18–0·38	115
Ukraine, deficient		<1·35	116
Armenian S.S.R., chestnut	4·6		117
Dark chestnut	5·3		117
Mountain-forest brown clay	6·7		117
Loamy chernozem	3·1		117
Brown sandy loam	3·2		117
Brown loam	3·0		117
Dark brown loam	2·2		117
Brown soddy mountain meadow	3·5		117
Alpine, brown sandy loam	4·5		117
Alpine, dark brown loam	4·3		117
Chuya valley	12–15	0·6–1·5	118
Tyan-Shan	2·50	0·06	119
Karshi steppe, irrigated light sierozems	8·5–13·0	0·14–0·45	120
Karshi steppe, non-irrigated light sierozems	5–10	0·12–0·32	120
Sevan Basin	1·6–6·72		121
Tien-Shan	6·2–11·9	0·3–1·3	122
Southern Ukraine	3–35	0·2–2·4	123
Flood-plain permafrost	8·22	3·04	124
Khar'kov region, strongly deficient in cobalt		<1·5	125
Deficient in cobalt		1·5–2	125
With a minimum of cobalt		2–3	125
With a reserve of cobalt		>3	125
Lugansk region, chernozems on sand and sandy loam		0·05–2·4	126
Calcareous chernozems and chernozems on elluvial shale		2·1–3·3	126
Common leached chernozems of northern region		2–4·5	126
Thick and transitional chernozems of southern region		3·8–4·7	126
Meadow and meadow-chernozem		4·9–5·7	126
Grey forest and soddy-meadow		0·07–0·65	127
Belorussian, soddy podzolic loams		0·10–0·58	128
Soddy podzolic sandy loams		0·25–1·0	128

Table 1   Cobalt Content of Soils (cont'd)

Soil	Cobalt (mg kg^{-1}) Total	Available	Ref.
U.S.S.R. (cont'd)			
Soddy podzolic loam on morainic loams		0·30–1·10	128
Soddy podzolic loam on loess		0·30–1·05	128
Soddy podzolic loam on lacustrine–glacial deposits		0·50–1·0	128
Soddy carbonate loam		1·05–2·20	128
Soddy podzolic marshy		0·30–1·25	128
Soddy marsh soils		0·62–3·0	128
Peat-marsh soils of lower peat type		1·4–3·2	128
Peat-marsh soils of transitional and upper peat type		0·5–2·4	128
Azerbaidzhan, deficient		0·64–1·84	129
Sheki-Zakataly region	5·8–8·0		130
Aktyubinsk region, solonets	3·05–6·33		131
Ivanov region, peats	0–2		132
Kazakhstan, chernozem		1·3–2·0	133
Aktyubinsk district, deficient	9·1–12·6	0·5–1·4	134
European U.S.S.R.	10	1·6	135
Khar'kov district	21·6–30·1		136
Smolensk, sod podzolic sandy	10		137
Sandy loam	< 10		137
Podzolized brown		2·92–3·12	138
Kirghiz	5·5–15	0·3–2·3	139
Poltava		0·25–2·90	140
Strongly podzolized soddy		0·3	141
Sod-medium podzolized sandy		0·3–0·7	141
Sod-weakly podzolized sandy		0·7	141
Sod-carbonate leached		1·9	141
Kamchatka, peat-bog soils, deficient		0·6–0·7	142
Zaysky-Burien	8–10·3	0·92–1·54	143
Mountain-meadow and mountain chernozem	14–22		144
Dark chestnut	19–24		144
Light chestnut	10		144
Sierozem	8–13		144
Sandy loam, deficient		0·35	145
Sheki-Zakataly zone, heavy, alluvial-heavy, light loam	1·0–7·1		146
Medium-loamy, weakly solonetz, sandy alluvial	0·3–4·4		146
Fergana Valley, sierozems		0·001–0·003	147

Table 1   (cont'd)

Soil	Cobalt (mg kg^{-1}) Total	Available	Ref.
U.S.S.R. (cont'd)			
Grey–brown		0·0001–0·001	147
Meadow		0·001–0·01	147
Solonchak		0·0003–0·001	147
Kedabek region	8–20		148
Kazakhstan	3–26		149
Kuibyshev district	4–25	1·2–6·8	150
Peat	2·4–6·0	0·5–5·6	151
Ul'yanovsk region, chernozems	3·7–12·3	0·40–0·73	152
Dark grey forest	10·3	0·35	152
Grey and light grey, forest	6·3	0·56	152
Sod-carbonate	2·1	0·27	152
Far East	8–10		153
Sumy district	11–35	0–1·7	154
Belorussian S.S.R., bog	3·3–7·88	0·2–0·72	155
Belorussian, bog	0·93–7·99	0·16–2·04	155
Belorussian, bog	0·65–7·74	0·28–2·92	155
Belorussian, bog	0·44–4·51	0·28–3·55	155
Chuvash S.S.R.	1·2–17·15		156
Vladimir region	1·7–17·2	1–10·5	157
Blagoveshensk Agricultural Institute		1–2	158
Dagestan plain	13·3–50·1	1·4–6·3	159
Tadzhikistan	8–16	0·7–2·1	160
Carbonate chernozem max.	1·173	0·191	161
Calcareous chestnut soil max.	0·708	0·098	161
Toguz-Torous basin	10·9–14·1	1·2–2·2	162
Chernozems	11·6		163
Central Chernozem belt:			
Podzolized	9·3–10·8		164
Leached	8·3–14·5		164
Typical	10·5–16·1		164
Common	11·2–15·7		164
Southern	11·7–16·2		164
Belorussia, swamp		0·21–3·12	165
Udmurtia			
Sod-weakly podzolic, loamy		2·26	166
Sod-strongly podzolic, loamy		1·04	166
Sandy loam and sandy		0·87	166
Grey wooded podzolized		2·77	166
Sod calcareous		2·50	166
Flood plain, semi-hydromorphic and hydromorphic		0·2–2·4	166

Table 1　Cobalt Content of Soils (cont'd)

Soil	Cobalt (mg kg^{-1})		Ref.
	Total	Available	
U.S.S.R. (cont'd)			
Caucasus, mining areas	7–22		167
Moscow region, flood plain	4·7–12·9		168
Kazakhstan, chestnut and meadow		0·86–0·87	169
Chernozems		1·8	169
Loriiskaya Steppe, chernozems	7·2–12·1	2·3–4·9	170
Yugoslavia			
Fertile	2·8		1
Peat		0·07	171
Zambia			
Dambo soil and stream sediments			
Mineralized area	20–160		172
Unmineralized area	3–12		172

Varying figures have been reported for the minimum quantity of total or available cobalt required in the soil to prevent deficiency diseases of cattle and sheep. Several workers have accepted 0·3 mg available Co kg^{-1} of soil as the minimum for good animal nutrition.[24,145,174] Other investigators place the minimum higher, at about 1·35,[116,134] 1·5,[90,94] 1·5–2,[125] 1·66,[110] 1·84,[129] and <2 mg available Co kg^{-1} soil.[158,164] A value of 2·5 mg total Co kg^{-1} of soil has been found by some workers to be the minimum for avoiding cobalt deficiency in ruminants;[5,6,73] another paper reported a value of 2·8.[78] One worker found that a soil containing 8–10 mg total Co kg^{-1} soil was deficient.[153]

It has been reported that in vineyard soils a content of 1·6 mg kg^{-1} total cobalt is adequate, but below 0·3 mg kg^{-1} the plants do suffer.[85]

Numerous papers have discussed the relation between cobalt content and soil type or parent material. Soils formed from granite in the United States[65] and Bulgaria[186] have a low cobalt content, but in Japan cobalt is higher in granitic and volcanic ash soils than in alluvial soils.[187] In the eastern United States, cobalt is lowest in soils with podzol morphology, and highest in alluvial soils and regosols.[188] Russian workers[73,189] have reported that sandy soils are generally deficient in cobalt, whereas heavier clay soils usually contain adequate amounts; highest quantities of cobalt occur in the lower horizons of lake clay deposits, moraine loams, and the alluvial horizons of silt loams. The concentration of cobalt in sandy loam is higher than in fluvoglacial and alluvial sands.[190] In the tilled horizon, soddy-meadow soils had the highest cobalt content, followed in decreasing order by grey forest soils, leached chernozems, soddy-podzolized soils, and peat-bog soils.[191]

In banana plantations along the Ivory Coast, the soils over basic rocks have the highest cobalt contents.[192] In Scotland, high levels of extractable cobalt occur in gley horizons of very poorly drained soils.[193] Saline soils of central Tien-Shan have a low cobalt content,[122] but saline soils of the Bashkir Trans-Ural region contain a high amount of cobalt.[194]

Conflicting results have been reported on the correlation between cobalt and iron in soils. A positive correlation between total cobalt and iron appears to exist in Gujarat soils of India,[26] Angola,[195] U.S.S.R.,[196] and Cuba.[14] No correlation was found in Mirzapur soils of India,[197] and high-iron soils of Australia[198] and Mali[199] are cobalt-deficient.

In many soils the cobalt content is higher in the surface layers.[15,25,82,84,94,123,200-203] Other soils, however, have more cobalt in the deeper layers.[29,73,139,204,205] In Dahomey, cobalt varies with depth in the same way as does the clay content,[206] and in basic soils of Tadzhikistan, cobalt content could not be correlated with horizons.[160]

The relation between humus and cobalt content of soil has been studied by many workers. Most have reported a higher content of cobalt in humus layers than in other horizons, or an accumulation of this element in organic matter.[52,96,154,164-167,207-211] It has also been found, however, that in some soils there was no accumulation of cobalt in the humus layer.[212,213]

All investigations reported on peat soils indicate that their cobalt content is low, and that deficiency diseases of livestock are common in such areas. [142,155,201,214,215] This may seem at variance with the results of most workers, cited in the preceding paragraph, that cobalt accumulates in humus layers. It is probable, however, that cobalt in peat soils would tend to be leached out from a medium which is usually saturated with water for most of the year. It has been observed that the botanical composition of peat influences the amount of cobalt in such soils; cobalt is high in woody peat components and low in various grass components.[155]

The effect of soil pH on cobalt availability has been noted in a few publications.[27,30,76,100] In soils of Uttar Pradesh, India, available cobalt increased with decreasing pH, but in Ukrainian soils higher available cobalt was found in neutral and alkaline conditions. In the Chkalov region of Russian and the Gujarat district of India, cobalt content was not related to the pH of the soil.

Linked to the effect of soil pH is the influence of lime additions on the availability of cobalt. Not unexpectedly, in view of the insolubility of cobalt carbonate in water, liming converts cobalt compounds in the soil to a less available state.[216-218] Therefore the full beneficial effects of liming in areas where soil cobalt is low, can only be realized when additional cobalt is provided for the plants and animals in such areas.

It has been found in both Finland[219] and the U.S.S.R.[220,221] that the finer soil textures tend to be richer in cobalt. The latter is concentrated mainly in silt-like fractions of 0·01–0·05 mm.

Several papers have reported that cultivated soils contained greater quantities of cobalt than uncultivated soils of similar type.[89,220,221]

An interesting observation has been made by Russian workers that cobalt content is higher in river flood-plain soils than in watershed and river terrace soils.[222,223] This confirms an earlier determination of a high cobalt content, 100–130 mg kg^{-1}, in Nile sediment.[89] The latter has supported continuous farming for over 5000 years without exhausting the land.

Radio isotopes have been used to study the influence of the vegetative cover on the migration of cobalt in soil. Vertical migration of cobalt was little influenced by vegetation; the loss of cobalt by leaching and plant assimilation was very low, nearly all the assimilated cobalt remaining in the parts of the roots which were in contact with the soil particles.[224] Likewise, vegetative cover was without influence on the horizontal migration of cobalt, which remains in the soil at the point of its introduction.[225]

A number of papers cover various aspects of the binding or sorption of cobalt in soil. The subject was reviewed briefly some years ago.[226] It has been demonstrated that cobaltiferous clays could be synthesized by incorporating cobalt into octahedral positions from dilute solutions.[227] The sorption of cobalt by three soils in the pH range 3–9 was found to be constant, but was very low at pH < 3; sorption depended on the presence of organic substances capable of forming chelates with cobalt.[228] It has been stated that cobalt sorbed by illites and kaolinite is not so readily available for plants as that fixed by montmorillonite.[229] One worker has found that the higher the pH the greater is the amount of cobalt which binds with organic soil compounds.[230] Another study indicated that cobalt was not as firmly bound to soil as was copper.[231] In some peats, a significant portion of cobalt existed in a free form, associated with organic substances in a colloidal complex.[132] In permanent pastures in New Zealand it was found that fertilizer addition is the major contributor to changes in soil cobalt, and the amount sorbed appears to depend on the organic matter of the soil, and on pH.[232]

The cobalt and vitamin B_{12} contents in clays and sediments of stagnant water reservoirs have been investigated, and a close relation between them was found.[233–235]

Papers have appeared on the relation between the cobalt content of soil and the occurrence of goiter.[90,169] Low available cobalt is positively correlated with the incidence of goiter; areas of high cobalt, even when low in iodine, appear to be relatively goiter-free.

It has been reported that cobalt added to soil containing organic matter accelerated the decomposition of the latter, and resulted in increased nitrogen fixation and improved availability of phosphorus.[236]

The suggestion has been made that the measurement of cobalt adsorbed by a soil can be used to determine the quantity of exchangeable bases, and the total cation-exchange capacity of that soil.[237]

A significant correlation has been obtained on seventy-one soils in the southeastern United States, between extractable cobalt from the A_1 horizon, and cobalt content in the leaves of black gum trees growing on such soils.[238]

The mineralogy and chemistry of soil cobalt has been summarized.[239] Cobalt, with other trace metals, in soils, plants, and animals, has been discussed in an extensive review.[240] Cobalt is included in a monograph on trace elements and colloids in soils.[241]

The proportion of samples in a sandy, grassland soil in Holland having an unacceptably low cobalt content, 0.3 mg kg^{-1}, increased from 1966–68 to 1971–73.[242]

In concluding this chapter, it will be noted that the number of references from Russian literature greatly exceeds that from the rest of the world combined. This will not be surprising to those who are acquainted with the leading role which Russia has long-occupied in soil science.

The absence of any published work on soil cobalt from China is regrettable. In view of the importance of food to this vast country, it might have been expected that the early use of cobalt compounds for imparting colors to pottery and porcelain would have been extended to studies on the occurrence and role of this element in Chinese agriculture.

References

1. Bertrand, D. and Mokragnatz, M. (1922). *Bull. Soc. Chim.* **31**, 1330–3.
2. McHargue, J. S. (1925). *J. Agric. Res.* **30**, 193–6.
3. Underwood, E. J. and Harvey, R. J. (1938). *Aust. Vet. J.* **14**, 183–9.
4. Nicolls, K. D. and Honeysett, J. L. (1964). *Aust. J. Agric. Res.* **15**, 368–76.
5. Pereira, J. A. A., Da Silva, D. J., and Braga, J. M. (1971). *Experientiae* **12**, 155–88; (1972). *C.A.* **76**, 125761.
6. Dantas, H. da S. (1971). *Pesqui Agropecuar. Brasil. Ser. Agron.* **6**, 23–6; (1973). *C.A.* **78**, 96477.
7. Patterson, J. B. E. (1938). *Emp. J. Exp. Agric.* **6**, 262–7.
8. Patterson, J. B. E. (1946). *Nature, Lond.* **157**, 555.
9. Corner, H. H. and Smith, A. M. (1938). *Biochem. J.* **32**, 1800–5.
10. Wright, J. R., Levick, R., and Atkinson, H. J. (1955). *Soil Sci. Soc. Am. Proc.* **19**, 340–4.
11. Wright, J. R. and Lawton, K. (1954). *Soil Sci.* **77**, 95–105.
12. Rana, S. K. and Oullette, G. J. (1967). *Can. J. Soil Sci.* **47**, 83–8.
13. Frank, R., Ishida, K., and Suda, P. (1976). *Can. J. Soil Sci.* **56**, 181–96.

14. Szoke, K. (1975). *Agrokem Talajtan* **24**, 445–50; (1976). *C.A.* **84**, 178768.
15. Nielsen, J. D. (1969). *Tidsskr. Planteavl.* **72**, 610–17; (1969). *C.A.* **71**, 48745.
16. Swain, D. J. (1955). Commonw. Bur. Soil Sci. Tech. Commun. 48.
17. Langenbeck, W. (1959). *Naturwissenschaften* **46**, 324; (1960). *C.A.* **54**, 6003.
18. Coppenet, M. and Calvez, J. (1967). *C.R. Hebd. Séances Acad. Agric. Fr.* **53**, 939–47; (1968). *C.A.* **68**, 68042.
19. Riehm, H. and Baron, H. (1953). *Landw. Forsch.* **5**, 145–58; (1954). *C.A.* **48**, 2958.
20. Riehm, H. and Scholl, W. (1957). *Landw. Forsch., Sonderh.* No. 9, 123–9; (1957). *C.A.* **51**, 14179.
21. Wehrman, J. (1955). *Pl. Soil* **6**, 61–83; (1955). *C.A.* **49**, 7168.
22. Seiffert, H. H. and Wehrman, J. (1957). *Z. PflErnähr. Düng. Bodenk.* **79**, 142–54; (1959). *C.A.* **53**, 7477.
23. t'Hart, M. L. and Deijs, W. B. (1952). *Phorphorsäure* **12**, 370–9; (1953). *C.A.* **47**, 4020.
24. Henkens, C. H. (1959). *Landbouwvoorlichting* **16**, 642–51; (1960). *C.A.* **54**, 14543.
25. Iyer, J. G. and Satyanaryan, Y. (1959). *J. Biol. Sci.* **2**, 110–15.
26. Reddy, K. G. and Mehta, B. V. (1961). *Soil Sci.* **92**, 274–80.
27. Reddy, K. G. and Mehta, B. V. (1962). *J. Indian Soc. Soil Sci.* **10**, 167–73.
28. Badhe, N. N. and Zende, G. K. (1962). *Indian J. Agron.* **6**, 304–10.
29. Mehta, B. V., Reddy, G. R., Nair, G. K., Ghandi, S. C., Neelkantan, V., and Reddy, K. G. (1964). *J. Indian Soc. Soil Sci.* **12**, 329–42.
30. Singh, S. and Singh, B. (1966). *J. Indian Soc. Soil Sci.* **14**, 177–81.
31. Cambi, G. (1949). *Annali Sper. Agra.* **3**, 963–73; (1953). *C.A.* **47**, 5595.
32. Giovannini, E., Usai, R., and Dore, G. (1954). *Studi Sassar., Sez.* 111, **2**, 60–79; (1956). *C.A.* **50**, 8114.
33. Bottini, E., Sapetti, C., and di Lavriano, E. M. (1959). *Annali Sper. Agra.* **13**, 499–522; (1959). *C.A.* **53**, 22299.
34. Carloni, L. (1963). *Chim. Ind.* **45**, 1084–6; (1964). *C.A.* **60**, 12614.
35. Garoia, V., Goldberg, F. L., and Amelotti, G. (1968). *Agricoltore Ital.* **68**, 67–80; (1969). *C.A.* **70**, 2758.
36. Chamberlain, G. T. (1959). *E. African Agric. J.* **25**, 121–5.
37. Bekhtur, U. (1960). Mikroelementy v Pochvakh, Vodakh i Organizmakh Vost. Sibiri i Dal'nego Vostoka i ikh Rol v Zhizni Rast., Zhivotn. i Cheloveka, Akad. Nauk S.S.S.R., Sibersk. Otdel., Trudȳ. Pervoi Konf., Ulan-Ude, 44–7; (1963). *C.A.* **59**, 10715.
38. Zhamsran, Zh. and Amgalan, Zh. (1972). Mikroelem. Biosfere Ikh Primen. Sel'. Khoz. Med. Sib. Dal'nego Vostoka, Dokl. Sib. Konf. 4th, 48–52; (1975). *C.A.* **83**, 95425.
39. Battseren, Ts., Gard'khu, Zh., Badarch, D., and Losolmaa, Zh. (1972). Mikroelem. Biosfere Ikh Primen. Sel'. Khoz. Med. Sib. Dal'nego Vostoka, Dokl. Sib. Konf. 4th, 94–7; (1975). *C.A.* **82**, 96919.
40. McNaught, K. J. (1938). *New Zealand J. Sci. Technol.* **20A**, 14–30.
41. Kidson, E. B. (1937). *New Zealand J. Sci. Technol.* **18**, 694–707.
42. Kidson, E. B. (1938). *J. Soc. Chem. Ind.* **57**, 95–6.
43. Wahhab, A. and Bhatti, H. M. (1958). *Soil Sci.* **86**, 319–23.
44. Khan, M. A. and Dubash, E. (1958). *Pakistan J. Sci. Res.* **10**, 123–8.
45. Strzemski, M. and Kabata, A. (1956). *Medycyna Wet.* **12**, 86; (1958). *C.A.* **52**, 630.

46. Basznski, T. (1958). *Acta Agrobotan.* **7**, 131–42; (1958). *C.A.* **52**, 18984.
47. Kabata, A. (1954). *Rocz. Glebozn.* **3**, 323–32; (1957). *C.A.* **51**, 11630.
48. Kabata, A. (1955). *Rocz. Nauk Roln., Ser. A.* **70**, 609–15; (1955). *C.A.* **49**, 16293.
49. Kabata-Pendias, A. (1968). *Rocz. Nauk Roln., Ser. A.* **94**, 567–83; (1969). *C.A.* **71**, 90280.
50. Roszyk, E. (1968). *Zesz. Nauk Wyzsz. Szk. Roln. Wroclaw.* **24**, 7–29; (1969). *C.A.* **70**, 28035.
51. Greinert, H. (1968). *Rocz. Glebozn.* **18**, 467–86; (1969). *C.A.* **70**, 2776.
52. Galczynska, B. and Piotrowska, M. (1972). *Pamiet. Pulawski* No. 53, 99–116; (1973). *C.A.* **79**, 114404.
53. Bajescu, I. and Bajescu, N. (1958). Acad. Rep. Populaire Romine, Inst. Cercetari Agron., 115–25; (1959). *C.A.* **53**, 15436.
54. Abadi, V. and Afusoaie, D. (1960). *Acad. Rep. Populaire Romine, Filiala Iasi, Studii Cercetari Stiint., Chim.* **11**, 263–70; (1962). *C.A.* **57**, 1292.
55. Suciu, T. and Ivanof, L. (1963). *Acad. Rep. Populaire Romine, Filiala Cluj, Studii Cercetari Agron.* **14**, 79–81; (1965). *C.A.* **63**, 13982.
56. Abadi, V., Dumitrescu, M., Ivanov, N., Murariu, T., and Sabliovschi, V. (1973). *An. Stiint. Univ. "Al. I. Cuza" Iasi, Sect. Ic* **19**, 97–104; (1973). *C.A.* **79**, 125069.
57. Burriel, F. and Gallego, R. (1952). *An. Edafol. Fisiol. Veg.* **11**, 569–600; (1953). *C.A.* **47**, 5595.
58. Ch'en, C. T., Wei, C. T., Yeh, C. P., and Cheng, J. M. (1966). *Chung Kuo Nung Yeh Hua Hsueh Hui Chih* **4**, 19–24; (1967). *C.A.* **66**, 75319.
59. Long, M. I. E. and Frederiksen, S. (1970). *Z. PflErnähr. Düng. Bodenk.* **126**, 238–44; (1971). *C.A.* **74**, 86945.
60. Seay, W. A. and DeMumbrum, L. E. (1958). *Agron. J.* **50**, 237–40.
61. Johnson, F. R. and Graham, E. R. (1952). *Missouri Agric. Exp. Stn Res. Bull.* 517.
62. Hill, A. C., Toth, S. J., and Bear, F. E. (1953). *Soil Sci.* **76**, 273–84.
63. Slater, S. C., Holmes, R. S., and Byers, H. G. (1937). U.S. Dept. Agric. Tech. Bull. 552.
64. Kubota, J. and Lazar, V. A. (1958). *Soil Sci.* **86**, 262–8.
65. Kubota, J. (1964). *Soil Sci. Soc. Am. Proc.* 1964, **28**, 246–51.
66. Peive, J. and Aizupiete, I. P. (1949). *Latv. PSR Zinät. Akad. Vest.* No. 5, 19–27; (1953). *C.A.* **47**, 10161.
67. Berzin, Y. M. (1950). *Agrobiologiya* **6**, 111–20; (1951). *C.A.* **45**, 5839.
68. Peive, J. (1950). Mikroelementy v Zhizni Rast. i Zhivotn., Akad. Nauk S.S.S.R., Trudy Konf. Mikroelem., 466–72; (1955). *C.A.* **49**, 550.
69. Katalymov, M. V. and Shirshov, A. A. (1955). *Dokl. Akad. Nauk S.S.S.R.* **101**, 955–7; (1955). *C.A.* **49**, 12615.
70. Malyuga, D. P. and Makarova, A. I. (1954). *Dokl. Akad. Nauk S.S.S.R.* **98**, 811–13; (1955). *C.A.* **49**, 4917.
71. Ronov, A. B., Malyuga, D. P., and Makarova, A. I. (1955). *Dokl. Akad. Nauk S.S.S.R.* **105**, 129–32; (1956). *C.A.* **50**, 7687.
72. Vlasyuk, P. A. (1955). Mikroelementy v Sel'. Khoz. i Med. Akad. Nauk Latv. S.S.R., Otdel. Biol. Nauk, Trudy Vsesoyuz. Soveshch. Riga, 97–103; (1959). *C.A.* **53**, 10618.
73. Peive, J. (1955). Mikroelementy v Sel'. Khoz. i Med. Akad. Nauk Latv. S.S.R., Otdel. Biol. Nauk, Trudy Vsesoyuz. Soveshch. Riga, 89–95; (1959). *C.A.* **53**, 10625.

74. Mikhel'son, Kh. K. (1955). Mikroelementy v Sel'. Khoz. i Med., Akad. Nauk
    Latv. S.S.R., Otdel. Biol. Nauk, Trudy Vsesoyuz. Soveshch. Riga, 297–304;
    (1959). C.A. **53**, 9545.
75. Peive, J. (1956–7). *Mikroelementy v Rastenievod., Akad. Nauk Latv. S.S.R., Inst.*
    Biol., Trudy Lab. Biokhim. Pochv. i Mikroelement. **9**, 5–43; (1961). C.A. **55**,
    20278.
76. Shifrina, P. A. and Batalin, A. Kh. (1957). *Vest. Chkalov. Oblast. Otdel.*
    Vsesoyuz. Khim. Obshchestva im D. I. Mendeleeva **7**, 11–13; (1961). C.A. **55**,
    2974.
77. Batalin, A. Kh., Bogdanova, E. S., Popova, A. A., Sadovskaya, L. V.,
    Filimonova, Z. G., Khmelevskaya, N. A., and Shtark, P. A. (1957). *Vest.*
    Chkalovsk. Khim. Obshchestva im D. I. Mendeleeva **7**, 7–9; (1961). C.A. **55**, 3898.
78. Narkevieius, J. (1959). *Trudy Litovsk. Vet. Akad.* **5**, 121–6; (1961). C.A. **55**,
    11728.
79. Egorova, T. K. (1959). *Udobrenie Urozhai* **4**, 32–5; (1960). C.A. **54**, 25458.
80. Grazhdan, P. E. (1959). *Izv. Akad. Nauk Turkmen. S.S.R.* **1**, 58–65; (1959).
    C.A. **53**, 13472.
81. Yakushevskaya, I. V. (1960). *Pochvovedenie* **6**, 92–6; (1960). C.A. **54**, 21577.
82. Lebedev, B. A. and Karavaev, V. N. (1960). Materialy po Izucheniyu Pochv
    Urala i Povolgh'ya Ufa. Sb., 178–82; (1962). C.A. **57**, 15533.
83. Dobrovol'skii, G. V. and Yakushevskaya, I. V. (1961). Rol Mikroelementov v
    Sel'. Khoz., Tr. 2-go (Vtorogo) Mezhvuz. Soveshch. po Mikroelementam, 32–6;
    (1962). C.A. **57**, 10251.
84. Toikka, M. A. (1961). Rol Mikroelementov v Sel'. Khoz., Tr. 2-go (Vtorogo)
    Mezhvuz. Soveshch. po Mikroelementam, 42–7; (1962). C.A. **57**, 8934.
85. Gadzhiev, D. M. (1961). Mikroelementy i Estestv. Radioaktivn. Pochv.
    Rostovsk. Gos. Univ., Materialy 3-go (Tret'ego) Mezhvuz. Soveshch., 186–8;
    (1963). C.A. **59**, 13295.
86. Rudin, V. D. (1962). *Trudy Stavropol'. Sel'.-khoz. Inst.* **12**, 21–7; (1964). C.A. **60**,
    4723.
87. Yur'eva, V. I. (1962). *Trudy Troitsk. Vet. Inst* **8**, 20–5; (1964). C.A. **60**, 11046.
88. Lukashev, K. I. and Petukhova, N. N. (1962). *Dokl. Akad. Nauk Beloruss. S.S.R.*
    6, 448–52; (1962). C.A. **57**, 14204.
89. Yakubov, A. M., Vel'gorskaya, N. N., and Ugenbaeva, M. A. (1962). *Vest. Sel'.-*
    khoz. Nauki, Min. Sel'. Khoz. Kaz. S.S.R. **5**, 43–54; (1962). C.A. **57**, 15538.
90. Gurevich, G. P. and Malyutina, L. I. (1962). *Trudy Vladivostoksk. Nauchn.-*
    Issled. Inst. Epidemiol., Mikrobiol. Gig., Sb. **2**, 211–13; (1963). C.A. **59**, 14519.
91. Purtova, A. T., D'yakova, Z. Z., and Zyuzina, V. G. (1962). Mikroelementy v
    Vost. Sibiri i na Dal'n. Vost., Inform. Byull. Koordinats. Komis. po
    Mikroelementam dlya Sibiri i Dal'n. Vost., 12–14; (1963). C.A. **59**, 15891.
92. Mirzaeva. K. Kh. (1963). Mikroelementy v Sel'.-khoz. i Med. Sb., 398–400;
    (1965). C.A. **62**, 13792.
93. Bel'gorskaya, N. N. and Yakubov, A. M. (1963). Mikroelementy v Sel'.-khoz. i
    Med. Sb., 435–41; (1965). C.A. **62**, 13792.
94. Lobanova, Z. I. (1963). Mikroelementy v Zhizni Rast., Zhivotn. i Cheloveka,
    Akad. Nauk Ukr. S.S.R., Inst. Fiziol. Rast., Tr. Koordinats. Sovesheb, 145–52;
    (1966). C.A. **64**, 8874.
95. Okhrimenko, M. F. (1963). Mikroelementy v Zhizni Rast., Zhivotn. i Cheloveka,
    Akad. Nauk S.S.R., Inst. Fiziol. Rast., Tr. Koordinats. Soveshch., 206–9;
    (1966). C.A. **64**, 8877.

96. Sedlukko, N. Ya. and Kondyukova, A. Kh. (1964). Pochv. Issled. i Rats. Ispol'z. Zemel Sb., 168–75; (1965). *C.A.* **63**, 8993.
97. Tonkonozhenko, E. V. (1964). *Trudȳ Kubansk. Sel'.-khoz. Inst.* **9**, 180–7; (1965). *C.A.* **63**, 8998.
98. Madanov, P. V., Voikin, L. M., and Evstigneeva, L. P. (1964). Okal'turivanie i Rats. Ispol'z. Pochv. i Udobreni Sb., 176–9; (1965). *C.A.* **63**, 13992.
99. Fat'yanov, A. S. (1964). Sb. Dokl. 111-ei (Tret'ei) Mezhob. Konf. Pochvovedov i Agrokhimikov Sred. Povolzh'ya i Yuzhn. Urala, Kuibyshev, 27–35; (1966). *C.A.* **64**, 10352.
100. Lobanova, Z. I. and Baranova, E. P. (1964). Soderzhanie Mikroelementov v Pochvakh Ukr. S.S.R. Sb., 127–43; (1966). *C.A.* **64**, 8875.
101. Kamynina, L. M. (1964). *Dokl. Rossinsk. Sel'.-khoz. Akad.* No. 99, 319–21; (1966). *C.A.* **64**, 2702.
102. Batmanova, O. Ya. (1965). *Trudȳ Leningrad. Sanit.-Gig. Med. Inst.* **81**, 46–9; (1967). *C.A.* **66**, 94272.
103. Shakuri, B. K. and Akhundova, G. G. (1965). *Izv. Akad. Nauk Azerb. S.S.R. Ser. Biol. Nauk* No. 2, 82–8; (1965). *C.A.* **63**, 10607.
104. Kazaryan, E. S., Asratyan, G. S., and Stepanyan, M. S. (1965). *Soobshch., Lab. Agrokhim., Akad. Nauk Arm. S.S.R.* No. 6, 20–30; (1966). *C.A.* **65**, 6236.
105. Muzaleva, L. D. and Pershina, E. F. (1965). *Uchen. Zap. Petrozavodsk. Gos. Univ.* **13**, 21–8; (1966). *C.A.* **65**, 19254.
106. Grivtsova, G. I. and Toikka, M. A. (1965). *Uchen. Zap. Petrozavodsk. Gos. Univ.* **13**, 29–35; (1966). *C.A.* **65**, 19253.
107. Gashimi, A. and Shakuri, B. K. (1965). *Dokl. Akad. Nauk Azerb. S.S.R.* **21**, 68–71; (1966). *C.A.* **64**, 10349.
108. Ben-Utyaeva, G. S. (1965). Mikroelementy v Sel'.-khoz. Sb., 318–21; (1966). *C.A.* **64**, 10352.
109. Ibragimova, V. I. (1965). Mikroelementy v Sel'.-khoz. Akad. Nauk S.S.R., Otdel. Khim.-Tekhnol. i Biol. Nauk, 146–51; (1966). *C.A.* **64**, 4799.
110. Shakuri, B. K. (1965). Mikroelementy v Sel'.-khoz. Akad. Nauk Uzbek. S.S.R., Otdel. Khim. Tekhnol. i Biol. Nauk, 333–41; (1966). *C.A.* **64**, 11799.
111. Kozyreva, G. F. and Rish, M. A. (1965). Mikroelementy v Sel'.-khoz. Akad. Nauk Uzbek. S.S.R., Otdel. Khim.-Tekhnol. i Biol. Nauk, 227–31; (1966). *C.A.* **64**, 11804.
112. Gyul'akhmedov, A. N. (1965). Mikroelementy v Sel'.-khoz. Sb., 402–8; (1966). *C.A.* **64**, 1309.
113. Sergeeva, A. G. (1965). *Izv. Kuibyshev. Sel'.-khoz. Inst.* **17**, 133–8; (1967). *C.A.* **66**, 10218.
114. Kazaryan, E. S. (1965). Sb. Dokl. Zakavkaz. Nauch.-Sess. Krupnomasshtabn. Pochv. Agrokhim. Kartir., 313–22; (1967). *C.A.* **66**, 94270.
115. Eyubov, I. Z. (1965). *Trudȳ Azerb. Nauch.-Issled. Vet. Inst.* **19**, 231–6; (1967). *C.A.* **67**, 21077.
116. Vlasyuk, P. A. (1966). *Pochvovedenie* No. 2, 17–21; (1966). *C.A.* **64**, 1334.
117. Kazaryan, E. S. and Airuni, G. A. (1966). *Izv. Sel'.-khoz. Nauk, Min. Sel'. Khoz. Arm. S.S.R.* Nos. 11–12, 29–39; (1968). *C.A.* **68**, 28788.
118. Openlender, I. V. (1966). Mikroelem. Zhivotnovod. Rastenievod. Akad. Nauk Kirg. S.S.R., 137–40; (1967). *C.A.* **66**, 64688.
119. Distanov, G. K., Baismakov, A., and Tokusheva, G. V. (1966). Mater. Mezhvuz. Nauch.-Teor, Konf. Nauch.-Ped. Rabot. Aspir. Vssyh. Ucheb. Zaved. Kirg. S.S.R. 1st Frunze, 160–1; (1967). *C.A.* **67**, 31935.

120. Dekhkankhodzhaeva, S. Kh. (1966). *Trudy Inst. Pochv. Uzbet. S.S.R.* No. 5, 352–8; (1967). *C.A.* **67**, 107721.
121. Kazaryan, E. S. and Airuni, G. A. (1967). *Izv. Sel'.-khoz. Nauk, Min. Sel'.Khoz. Arm. S.S.R.* **10**, 65–72; (1968). *C.A.* **68**, 28789.
122. Openlender, I. V. (1967). *Mikroelem. Zhivotnovod. Rastenievod. Akad. Nauk Kirg. S.S.R.* No. 6, 78–89; (1968). *C.A.* **69**, 51198.
123. Dobrolyubskii, O. K. (1967). *Mikroelem. Sel'. Khoz. Med.* No. 3, 1309; (1968). *C.A.* **69**, 105417.
124. Kurilyuk, T. T. (1967). *Mikroelem. Biosfere Ikh Primen. Sel'.-khoz. Med. Sib. Dal'nevost.*, 115–20; (1969). *C.A.* **70**, 2764.
125. Krupskii, N. K., Aleksandrova, A. M., and Lysenko, M. N. (1967). *Mikroelem. Biosfere Ikh Primen. Sel'. Khoz. Med. Sib. Dal'nego Vostoka*, 186–90; (1969). *C.A.* **70**, 46436.
126. Tsyganenko, O. Yu. (1967). *Puti Povysh. Plodorodiya Pochv.*, 157–60; (1969). *C.A.* **70**, 46440.
127. Andreevskaya, V. E. (1967). *Probl. Ekol.* **1**, 14–17; (1969). *C.A.* **70**, 86613.
128. Lupinovich, I. S. and Misnik, A. G. (1967). *Dokl. Akad. Nauk Beloruss. S.S.R.* **11**, 439–43; (1967). *C.A.* **67**, 10770.
129. Dzhafarov, Ya. M. and Gyul'akhmedov, A. N. (1968). Mater. Respub. Konf. Probl. "Mikroelem. Med. Zhivotnovod" 1st, 13–14; (1970). *C.A.* **73**, 55057.
130. Gyul'akhmedov, A. N. and Gadzhiev, F. M. (1968). Mater. Respub. Konf. Probl. "Mikroelem. Med. Zhivotnovod" 1st, 11–13; (1970). *C.A.* **73**, 55056.
131. Klimakhin, N. A. and Ivanov, A. P. (1968). *Melior Solontsov*, 177–81; (1970). *C.A.* **72**, 65917.
132. Chistyakov, N. N. and Babanova, M. S. (1968). Prir. Ivanov.Obl., 39–41; (1970). *C.A.* **72**, 65920.
133. Grabarov, P. G. and Solodnikova, E. A. (1968). Agrokhim. Kharakter. Pochv. S.S.S.R., Kaz. Chelyabinsk. Obl., 154–64; (1970). *C.A.* **72**, 30622.
134. Gafarova, K. M. (1968). *Vest. Sel'.-khoz.Nauki, Alma-Ata* **11**, 91–3; (1969). *C.A.* **70**, 19236.
135. Zyrin, N. G. (1968). *Agrokhimiya* **11**, 84–90; (1969). *C.A.* **70**, 3665.
136. Litvinenko, M. G. (1969). *Gig. Naselennykh Mest* **8**, 5–8; (1970). *C.A.* **73**, 48400.
137. Vasil'evskaya, V. D. and Tyuryukanov, A. N. (1969). Pochvy Dernovo-Podzol. Zony Ikh Ratsion. Ispol'z, 134–50; (1970). *C.A.* **72**, 54154.
138. Chanturiya, I. A. (1969). *Subtropiki Kul't.* No. 3, 156–9; (1970). *C.A.* **72**, 2599.
139. Mamytov, A. M. and Openlender, I. V. (1969). *Izv. Akad. Nauk Kirg. S.S.R.* No. 2, 36–41; (1970). *C.A.* **72**, 20904.
140. Krupskii, N. K., Aleksandrova, A. M., and Lysenko, M. N. (1969). *Mikroelem. Sel'. Khoz. Med.* No. 5, 132–9; (1970). *C.A.* **73**, 55058.
141. Krym, I. Ya. (1969). *Sev.-Zapad Evr. Chasti S.S.R.* No. 7, 105–6; (1971). *C.A.* **75**, 75388.
142. Struzhkina, T. M., Tuzhilina, I. B., Shripchenko, A. F., Vorontsova, L. I., and Appolonova, G. K. (1969). *Uchen. Zap. Dal'nevost. Gos. Univ.* **27**, 66–71; (1972). *C.A.* **77**, 18586.
143. Kostikov, D. N. (1969). Mater. Nauch. Konf. Blagoveshchensk. Sel'.-khoz. Inst. Sekts. Agron. 18th, 49–53; (1973). *C.A.* **79**, 77428.
144. Shlavitskaya, Z. I., Kvetkina, A. A., and Fedorin, Yu. V. (1970). *Trudy Inst. Pochv., Akad. Nauk Kaz. S.S.R.* **18**, 147–52; (1970). *C.A.* **73**, 76225.
145. Dulesova, K. N. and Simakova, T. V. (1970). *Dokl. TSKHA* No. 159, 141–3; (1971). *C.A.* **74**, 98950.

146. Kuliev, Sh. M. (1970). *Izv. Akad. Nauk Azerb. S.S.R.* (5–6), 73–7; (1971). *C.A.* **75**, 97625.
147. Mirzaeva, K. Kh. (1970). *Nauch. Trudȳ Tashkent. Gos. Univ.* No. 377, 311–17; (1972). *C.A.* **76**, 58146.
148. Malyuga, D. P. and Niyazov, A. Kh. (1971). *Izv. Akad. Nauk Azerb. S.S.R. Ser. Biol. Nauk* No. 2, 72–8; (1972). *C.A.* **76**, 13127.
149. Voronova, E. P. (1971). *Pochvovedenie* No. 9, 49–61; (1971). *C.A.* **75**, 150711.
150. Peterburgskii, A. V. and Prazdnikova, R. V. (1971). *Dokl. TSKHA* No. 169, 40–5; (1972). *C.A.* **76**, 125723.
151. Gladilovich, B. R., Antonova, G. G., Vard'ya, N. P., Drel, R. I., Kurbatova, R. I., Mikhailyk, E. V., and Khristoforova, L. I. (1971). *Zap. Leningrad. Sel'.-khoz. Inst.* **160**, 4–9; (1972). *C.A.* **76**, 58142.
152. Neklyudov, V. N., Zusmanovskii, A. G., Merkulov, N. N., Kornev, S. D., and Sominskaya, E. Z.(1971). *Khim. Sel'. Khoz.* **9**, 336–9; (1972). *C.A.* **76**, 84767.
153. Golov, V. I. (1971). Agrokhim. Kharakter. Pochv. S.S.S.R., Dal'nii Vostok, 152–69; (1972). *C.A.* **77**, 4207.
154. Krupskii, N. K., Aleksandrova, A. M., and Lysenko, M. N. (1972). *Visn. Sil.-kogospod. Naukȳ* No. 1, 59–62; (1972). *C.A.* **77**, 47182.
155. Vashkevich, L. F. (1972). *Vest. Akad. Navuk B.S.S.R.*, *Ser. Sel'skagaspad. Navuk* No. 3, 25–31; (1973). *C.A.* **78**, 3148.
156. Golubev, I. M., Romashova, V. G., and Merkulov, N. N. (1972). *Trudȳ Ul'yanov. Sel'.-khoz. Inst.* **17**, 56–62; (1973). *C.A.* **79**, 41300.
157. Gladilovich, B. R., Antonova, G. G., Vard'ya, N. P., Drel, R. I., Kurbatova, R. I., Mikhailyk, E. V., and Travitskaya, E. O. (1972). *Zap. Leningrad. Sel'.-khoz. Inst.* **200**, 24–31; (1974). *C.A.* **80**, 132036.
158. Kostikov, D. N. (1972). Mikroelem. Biosfere Ikh. Primen. Sel'. Khoz. Med. Sib. Dal'nego Vostoka, Dokl. Sib. Konf. 4th, 91–3; (1975). *C.A.* **82**, 96918.
159. Mursalova, M. G. and Mirzoeva, Z. A. (1973). *Agrokhimiya* No. 2, 115–17; (1973). *C.A.* **78**, 123138.
160. Zyranova, A. N. (1973). *Dokl. Akad. Nauk Tadzh. S.S.R.* **16**, 59–63; (1973). *C.A.* **79**, 65049.
161. Rudin, V. D. and Shipotovskii, V. V. (1973). *Trudȳ Stavrop. Skh. Inst.* **3**, 136–9; (1975). *C.A.* **82**, 3173.
162. Obukhova, Z. D. and Dorozhlina, A. F. (1973). *Mikroelem. Zhivotnovod. Rastenievod.* **12**, 87–92; (1975). *C.A.* **83**, 42041.
163. Kirilyuk, V. P., Rabinovich, I. Z., and Unguryan, V. G. (1973). *Trudȳ Kishinev. Sel'.-khoz. Inst.* **99**, 150–5; (1974). *C.A.* **80**, 94555.
164. Aderikhin, P. G. and Kopaeva, M. T. (1974). *Agrokhimiya* No. 2, 116–22; (1974). *C.A.* **81**, 2740.
165. Svirnovskii, L. Ya. (1974). *Vesti Akad. Navuk B.S.S.R.*, *Ser. Sel'skagaspad. Navuk* No. 1, 19–25; (1974). *C.A.* **81**, 2748.
166. Isaev, M. A., Kuznetsov, M. F., and Kovrigo, V. P. (1974). *Trudȳ Izhevsk. S-kh. Inst.* **23**, 37–44; (1976). *C.A.* **84**, 16064.
167. Gyul'akhmedov, A. N., Kuliev, Sh. M., and Gyandzhemekhr, A. V. (1974). *Trudȳ Azerb. Fil. Vses. O-va. Pochvovedov*, 48–51; (1976). *C.A.* **84**, 72936.
168. Verigina, K. V. and Korableva, L. I. (1974). Biol. Rol Mikroelem. Ikh Primen. Sel'. Khoz. Med., 191–206; (1975). *C.A.* **82**, 96927.
169. Chigrina, T. A. and Shumaev, V. D. (1974). *Trudȳ Nauchno-Issled. Inst. Kraev. Patol.*, Alma-Ata **26**, 84–91; (1975). *C.A.* **83**, 191767.
170. Khachatryan, A. S. (1975). *Izv. S-kh. Nauk* **18**, 44–9; (1975). *C.A.* **83**, 57300.

171. Furlan, J. and Stupar, J. (1967). *Zemlj. Biljka* **16**, 691–6; (1968). *C.A.* **68**, 113629.
172. Webb, J. S. and Tooms, J. S. (1958–9). *Trans. Instn Min. Metall., Part IV* **68**, 125–44.
173. Koval'skii, V. V., Mursaliev, A., and Gribovskaya, I. F. (1966). *Agrokhimiya* No. 1, 87–99; (1966). *C.A.* **64**, 11802.
174. Mitchell, R. L. (1972). *Atti Simp. Int. Agrochim.* **9**, 521–32.
175. Orlova, L. P. and Ivanov, D. N. (1974). *Pochvovedenie* No. 2, 31–6; (1974). *C.A.* **80**, 132051.
176. Gille, G. L. and Graham, E. R. (1971). *Soil Sci. Soc. Am. Proc.* **35**, 414–16.
177. Gyul'akhmedov, A. N. (1961). *Izv. Akad. Nauk Azerb. S.S.R., Ser. Biol. Med. Nauk* No. 7, 57–63; (1962). *C.A.* **57**, 17103.
178. Minami, K., Yasuda, T., and Araki, K. (1973). *Tokai Kinki Nogyo Shikenjo Kenkyu Hokoku* **25**, 48–56; (1974). *C.A.* **80**, 132095.
179. Bajescu, I. and Dinu, E. (1970). *An. Inst. Stud. Cercet. Pedol.* **38**, 77–97; (1972). *C.A.* **77**, 113102.
180. Lysenko, M. N. (1967). Puti Povysh. Plodorodiya Pochv., 153–6; (1968). *C.A.* **69**, 43049.
181. Barkan, Ya. G., Klimova, V. L., and Sukhoverkhova, M. D. (1972). Mikroelem. Biosfere Ikh Primen. Sel'. Khoz. Med. Sib. Dal'nego Vostoka, Doklady Sib. Konf. 4th; (1975). *C.A.* **82**, 85083.
182. Arinushkina, E. V. (1961). Mikroelementy i Estestv. Radioaktivn. Pochv. Rostovsk. Gos. Univ., Materially 3-go (Tret'ego) Mezhvuz. Soveshch., 249–56; (1963). *C.A.* **59**, 15887.
183. Hodgson, J. F. (1960). *Soil Sci. Soc. Am. Proc.* **24**, 165–8.
184. Reddy, K. G. and Mehta, B. V. (1961). *Indian J. Agric. Sci.* **31**, 261–5.
185. Kubota, J. (1965). *Soil Sci.* **99**, 166–74.
186. Stoyanov, D. V (1963). *Izv. Nauch.-Issled. Inst. Pochv. Agrotekhn "Nikola Pushkarov", Akad. Selskostopansk. Nauki Bulgar.* **6**, 93–100; (1964). *C.A.* **60**, 1062.
187. Ikeda, M. and Kurozumi, K. (1967). *Hiroshima Daigaku Suichikusan Gakribu Kiyo* **7**, 149–70; (1968). *C.A.* **69**, 18305.
188. Kubota, J. and Lazar, V. A. (1960). Trans. Int. Congr. Soil Sci. 1st., Madison, Wisc., (1960) Vol. 2, 134–41.
189. Vil'gusevich, I. P. and Bulgakov, N. P. (1960). *Pochvovedenie* 104–11; (1960). *C.A.* **54**, 14531.
190. Dobrovol'skii, V. V. (1963). *Nauch. Dokl. Vyssh. Shk., Biol. Nauki* 193–8; (1964). *C.A.* **60**, 1064.
191. Davydov, N. I. and Starobinets, K. S. (1966). *Trudy Altai. Sel'.-khoz. Inst.* No. 9, 45–51; (1967). *C.A.* **66**, 114933.
192. Dabin, B. and Leneuf, N. (1960). *Fruits* No. 3, 117–27; (1960). *C.A.* **54**, 15790.
193. Swaine, D. J. and Mitchell, R. L. (1960). *J. Soil Sci.* **11**, 347–68.
194. Amirkhanova, S. N. (1960). *Khim. Sel'. Khoz. Bashkirii, Ufa* No. 2, 111–16; (1961). *C.A.* **55**, 21438.
195. Fragoso, M. A. C. (1959). *Mems Jta Invest. Ultramar* No. 11, 1–238; (1961). *C.A.* **55**, 13739.
196. Vekilova, F. I. and Barovskaya, Yu. B. (1960). *Izv. Akad. Nauk Azerb. S.S.R., Ser. Geol.-Geograf. Nauk* No. 6, 51–63; (1961). *C.A.* **55**, 14782.
197. Singh, L. and Singh, S. (1973). *Proc. Indian Nat. Sci. Acad., Part B* **39**, 136–41; (1974). *C.A.* **81**, 76813.
198. McKenzie, R. M. (1959). *Aust. J. Agric. Res.* **10**, 52–7.

199. Peive, J. (1963). *Pochvovedenie* No. 11, 47–50; (1964). *C.A.* **60**, 7391.
200. Toikka, M. A. (1960). Voprosy̅ Pochv. Karel'sk. S.S.R., 25–42; (1963). *C.A.* **58**, 1875.
201. Chodan, J. (1962). *Rocz. Nauk Roln., Ser. F.* **75**, 545–62; (1963). *C.A.* **59**, 15889.
202. Toikka, M. A. (1964). *Uchen. Zap. Petrozavodsk. Gos. Univ.* **12**, 57–63; (1966). *C.A.* **65**, 7952.
203. Dzhangaliev, A. D., Uvarov, Yu. P., and Andreev, G. E. (1966). *Izv. Akad. Nauk Kaz. S.S.R., Ser. Biol.* 4, 13–18; (1967). *C.A.* **66**, 75295.
204. Dobrovol'skii, V. V. (1960). *Pochvovedenie* No. 2, 15–23; (1960). *C.A.* **54**, 14532.
205. Korataev, N. Ya., Azarina, V. A., Vologzhanina, T. V., and Protasova, L. A. (1960). *Trudy̅ Permsk. Sel'.-khoz. Inst.* **17**, 116–21; (1961). *C.A.* **55**, 25125.
206. Pinta, M. and Ollat, C. (1961). *Geochim. Cosmochim. Acta* **25**, 14–23; (1961). *C.A.* **55**, 23896.
207. Makitie, O. (1961). *Maatalouden Tutkimuskieskus Maantutkimuslaitos Agrogeol. Julkaisuja* No. 78; (1961). *C.A.* **55**, 25123.
208. Sillanpaa, M. (1962). *Maatalouden Tutkimuskieskus Maantutkimuslaitos Agrogeol. Julkaisuja* No. 81; (1962). *C.A.* **57**, 12926.
209. Graham, E. R. and Killion, D. D. (1962). *Soil Sci. Soc. Am. Proc.* **26**, 545–7.
210. Dobritskaya, Yu. I., Zhuravleva, E. G., Orlova, L. P., and Shirinskaya, M. G. (1962). Mikroelementy v Sel'. Khoz. i Med. Ukr. Nauch.-Issled. Inst. Fiziol. Rast. Akad. Nauk Ukr. S.S.R., Materialy 4-go (Chetvertogo) Vses. Soveshch. Kiev, 391–7; (1965). *C.A.* **63**, 8991.
211. Freiberga, G. (1970). *Latv. PSR Zinat. Akad. Vest.* No. 2, 116–21; (1970). *C.A.* **73**, 24377.
212. Sedlukho, N. Ya. (1961). *Trudy̅ Beloruss. Sel'.-khoz. Akad.* **34**, 104–8; (1963). *C.A.* **58**, 4993.
213. Gyori, D. (1962). *Magy. Tudom. Akad. Agrártud. Osztál. Közl.* **21**, 53–71; (1963). *C.A.* **59**, 3280.
214. Kabata, A. (1958). *Rocz. Nauk Roln., Ser. A.* **78**, 379–450; (1959). *C.A.* **53**, 16451.
215. Tokovoi, N. A. and Kopeikin, Yu. A. (1961). Trudy̅ Pervoi Sibirsk. Konf. Pochvovedov, Akad. Nauk S.S.S.R., Sibirsk. Otdel., 463–71; (1964). *C.A.* **60**, 9850.
216. Kedrov-Zikhman, O. K. (1960). Sb. Nauch. Tr. po Izvestkovaniyu Dernovopodzolist. Pochv., 17–33; (1963). *C.A.* **58**, 6154.
217. Gyul'akhmedov, A. N. (1961). *Trudy̅ Tashkentsk. Konf. po Mirnomu Izpol'z. At. Energii, Akad. Nauk Uzbek. S.S.R.* **3**, 491–7; (1962). *C.A.* **57**, 11508.
218. Pilipushko, V. N. and Tuev, N. A. (1962). *Vest. Leningrad. Univ. Biol.* No. 3, 105–11; (1973). *C.A.* **78**, 3220.
219. Vuorinen, J. (1960). *Maatalous Koetoim.* **14**, 24–32; (1962). *C.A.* **56**, 7733.
220. Gavrilova, A. N. (1966). *Agrokhimiya* No. 7, 93–105; (1966). *C.A.* **65**, 12809.
221. Pobedintseva, I. G. (1967). Geogr. Pochv. Geokhim. Landshaftov, 184–96; (1968). *C.A.* **68**, 48594.
222. Dobrovol'skii, G. V. and Yakushevskaya, I. V. (1960). *Vest. Moscov. Univ., Ser. VI* **15**, 57–70; (1961). *C.A.* **55**, 14782.
223. Vasil'evskaya, V. D. (1958). *Nauch. Dokl. Vys̅sh. Shk., Biol. Nauki* No. 3, 179–82; (1960). *C.A.* **54**, 6004.
224. Kulikov, N. V. (1965). *Trudy̅ Inst. Biol., Akad. Nauk S.S.S.R., Ural'sk. Filial* No. 45, 97–106; (1966). *C.A.* **65**, 4581.
225. Poryadkova, N. A. and Agafonova, S. F. (1965). *Trudy̅ Inst. Biol., Akad. Nauk S.S.S.R., Ural'sk. Filial* No. 45, 107–14; (1966). *C.A.* **65**, 4582.

226. Kabata, A. (1955). *Postępy Nauk Roln.* **2**, 58–60; (1959). *C.A.* **53**, 14394.
227. Caillère, S., Henin, S., and Esquevin, J. (1958). *Clay Miner. Bull.* **3**, 232–7; (1959). *C.A.* **53**, 9915.
228. Chebotina, M. Ya. and Titlyanova, A. A. (1965). *Trudy Inst. Biol., Akad. Nauk S.S.S.R., Ural'sk. Filial* No. 45, 85–9; (1966). *C.A.* **65**, 4581.
229. Kabata-Pendias, A. (1968). *Rocz. Glebozn.* **19**, 55–72; (1968). *C.A.* **69**, 105373.
230. Agapov, A. I. (1968). *Agrokhimiya* No. 4, 106–12; (1968). *C.A.* **69**, 18282.
231. Kanunnikova, N. A. (1968). *Mikroelem. Sib.* No. 6, 23–9; (1969). *C.A.* **71**, 29679.
232. Forbes, E. A. (1976). *New Zealand J. Agric. Res.* **19**, 153–64.
233. Letunova, S. V. (1958). *Dokl. Vses. Akad. Sel'.-khoz. Nauk, im V.I. Lenina*, **21**, 31–4; (1959). *C.A.* **53**, 4619.
234. Letunova, S. V. (1959). *Trudy Tsent.-Chernoz. Gos. Zapov. im Prof. V. V. Alekhina* No. 5, 372–6; (1961). *C.A.* **55**, 14761.
235. Letunova, S. V. (1960). *Izv. Akad. Nauk Azerb. S.S.R., Ser. Biol. Sel'.-khoz. Nauk* No. 6, 143–6; (1961). *C.A.* **55**, 22666.
236. Singh Verma, S. B. and Pathack, A. N. (1962). *Agra Univ. J. Res.* **11**, 97–103; (1963). *C.A.* **58**, 6143.
237. Johansson, A. (1961). *Soil Sci.* **91**, 364–8.
238. Alban, L. A. and Kubota, J. (1960). *Soil Sci. Soc. Am. Proc.* **24**, 183–5.
239. McKenzie, R. M. (1974). Trace Elem. Soil–Plant–Animal Syst., Proc. Jubilee Symp., 83–93.
240. Lisk. D. J. (1972). *Adv. Agron.* **24**, 267–325.
241. Ermolenko, N. F. (1973). "Trace Elements and Colloids in Soils", 2nd Edn. Wiley-Interscience, Chichester.
242. Lammers, H. W. (1975). *Bedrijfsontwikkeling* **6**, 611–13; (1975). *C.A.* **83**, 177220.

3 Cobalt in Fertilizers

The addition of cobalt salts to fertilizers, or to soil additives such as limestones, quickly followed the realization that many areas in the world are deficient in this element. A number of studies have been made on the cobalt content of fertilizers and limestones,[1-18] and the results are summarized in Table 2.

It will be observed that most fertilizing materials are rather low in cobalt. Some, however, such as farmyard manure and limestone, are applied at high rates of several tonnes or more, per hectare, and cobalt additions thereby become appreciable. The application of limestone averaging 4 mg kg^{-1} was found to markedly increase the cobalt content of New Zealand pasture.[9] In the United States, it has been calculated that cobalt removed from crops in five years would be about 0·635 g ha^{-1}, whereas one application of approximately 4·5 tonnes ha^{-1} of limestone during this period would put into the soil 3·6–22·7 g of cobalt.[6]

An excellent way to make cobalt additions to a soil, crop, or pasture is to mix a cobalt salt with the fertilizer or limestone which is applied regularly, thereby ensuring an even distribution of the small quantity of cobalt required. An application of 1–2 kg cobalt sulphate per hectare of pasture is usually effective for 3–5 years in preventing cobalt deficiency in cattle and sheep.

The form of the anion is not critical in top-dressing pastures; the carbonate, phosphate, and sulphate of cobalt give equivalent results.[19] Cobalt chloride and nitrate may also, of course, be used. Cobaltized superphosphate has been employed extensively, the ratio being about 1 kg cobalt salt to 100 kg of superphosphate.[20] Some workers have found the incorporation of cobalt compounds with limestone entirely satisfactory, but others have reported that the uptake of cobalt is depressed by the presence of lime. It must be recognized that in many soil and climatic conditions at least a part of the cobalt may be temporarily immobilized by lime. In choosing the form of cobalt compound, it should be remembered that both the carbonate and phosphate of cobalt are relatively insoluble in water. They become soluble, of course, under the chemical and biological activities of soil solutions; their relative water-insolubility provides a slow release of cobalt and a long-lasting effect which is frequently very desirable.

Table 2   Cobalt Content of Fertilizers

Fertilizer	Cobalt (mg kg^{-1})	Ref.
Ammonium sulphate, Britain	0	1
Cobalt tailings, U.S.S.R.	138–228	2
Copper–cobalt fertilizer, Germany	3000	3
Cyanamide, Yugoslavia	7	1
Farmyard manure, Britain	6	1
Farmyard manure, Germany	0·06	4
Farmyard manure, Poland	0·11–6·76	5
Farmyard manure, Yugoslavia	6	1
Lime, by-product of lead–zinc operations, U.S.A.	1–54	6
Lime, by-product of lime manufacture, U.S.A.	<1	6
Lime, sugar beet refuse, U.S.A.	<1	6
Lime, sugar beet refuse, Yugoslavia	1	1
Limestones, Brazil	tr.–46·9	7
Limestones, Britain	0·72–2·1	8
Limestones, New Zealand	4	9
Limestones, U.S.A.	<1–6	6
Nitrate of soda, Chile	2–6	10
Oyster shells, Canada	1	11
Phosphate, U.S.S.R.	1–54	12
Phosphate, U.S.S.R.	1000	13
Potassium chloride, Britain	1	1
Potassium salts, U.S.S.R.	9	13
Potassium sulphate, Britain	0	1
Sewage sludge, India	0·5–0·8	14
Sewage sludge, U.S.A.	8·5	15
Sodium nitrate, synthetic, Britain	0	1
Superphosphate, Australia	5	16
Superphosphate, Britain	4	1
Superphosphate, Yugoslavia	9	1
Thomas phosphate, Germany	3·3	17
Thomas phosphate, Germany	1·8–7·9	18
Thomas phosphate, U.S.S.R.	1–145	12

The technical grades of cobalt carbonate and phosphate contain about 48–49% cobalt, while the chloride, nitrate, and sulphate range from approximately 20–24% cobalt. The price is fairly closely related to the cobalt content, irrespective of the compound. For some pastures on hilly and rugged terrain, cobalt top-dressings have been applied either as aqueous solutions or powders by aerial spraying.[21]

A few miscellaneous observations on cobalt in fertilizers have been published. The solids produced in activated sludge treatment showed a two-fold, or greater, increase of cobalt as compared with sewage solids;

comparable increases occurred in septic tank treatment.[22] In Poland, manuring with straw was found to increase the cobalt content of upland soils.[23] It has been found in Russia, that grain-fed cattle assimilated 20–25% of the cobalt in their food; about 75% can therefore be recovered in the manure and litter.[24] When the optimum addition of cobalt is assumed to be 0.16 kg ha^{-1}, the best application of phosphorite is 50 kg P_2O_5 ha^{-1}, and since the phosphorite contains 20% P_2O_5, the minimum concentration of cobalt in phosphorites should be about 0.02% or 200 mg kg^{-1}.[25] When cobalt is added as heptahydrate sulphate during ammophos production, at pH 4–6, an insoluble cobalt ammonium phosphate, $CoNH_4PO_4.H_2O$, is formed.[26]

An extensive compilation of trace elements in fertilizers (prepared a few years ago and, unfortunately, now out of print), has a number of references to cobalt.[27] The values ranged from <0.001 mg kg^{-1} to >100 mg Co kg^{-1}, but the majority were about 1 mg kg^{-1}. The amounts of cobalt supplied by fertilizing have been reviewed in a paper on trace-element balance.[28]

It is of interest to note that a recent study of the aerial deposition of cobalt in Britain has indicated that this amounts to 0.15–0.41 mg m^{-2} per year. This is sufficient to satisfy most of the crop requirement; in cleaner regions of the world, deposition might be only one-tenth of this value.[29]

References

1. Stojkovska, A. and Cooke, G. W. (1958). *Chem. Ind.* Oct. 18, 1368.
2. Velikova, F. I. and Borovskaya, Y. B. (1959). *Izv. Akad. Nauk Azerb. S.S.R., Ser. Geol-Geograf. Nauk* No. 3, 73–4; (1959). *C.A.* **53**, 19257.
3. Scharrer, K., Kuhn, H., and Schaumloffel, E. (1960). *Landw. Forsch.* **13**, 34–42; (1960). *C.A.* **54**, 15797.
4. Scharrer, K. and Prun, H. (1956). *Landw. Forsch.* **8**, 182–206; (1956). *C.A.* **50**, 12378.
5. Mazur, T. (1972). *Rocz. Glebozn.* **23**, 305–13; (1973). *C.A.* **78**, 134997.
6. Chichilo, P. and Whittaker, C. W. (1961). *Agron. J.* **53**, 139–44.
7. Valadares, J. M. A. S., Bataglia, O. C., and Furlani, P. R. (1974). *Brazantia* **33**, 147–52; (1975). *C.A.* **83**, 8239.
8. Batey, T. and Williams, T. R. (1966). *Chem. Ind.* Dec. 31, 2199–3000.
9. Dixon, J. K. and Kidson, E. B. (1940). *New Zealand J. Sci. Technol.* **22**, 1–6.
10. Young, R. S. (1951). *Research* **4**, 392.
11. Young, R. S. (1960). *J. Agric. Food Chem.* **8**, 485–6.
12. Katalymov, M. V. and Shirshov, A. A. (1955). *Dokl. Akad. Nauk S.S.S.R.* **101**, 955–7; (1955). *C.A.* **49**, 12615.
13. Kropachev, A. M. and Kropacheva, T. S. (1965). *Vest. Sel'.-khoz. Nauki, Min. Sel'. Khoz. S.S.S.R.* **10**, 68–9; (1966). *C.A.* **65**, 4588.
14. Rao, S. S. and Srinath, E. G. (1960). *Naturwissenschaften* **47**, 518; (1961). *C.A.* **55**, 10760.
15. Clark, L. J. and Hill, W. L. (1958). *J. Ass. Off. Agric. Chem.* **41**, 631–7.

16. Bryan, W. W., Thorne, P. M., and Andrew, C. S. (1960). *J. Aust. Inst. Agric. Sci.* **26**, 273–5.
17. Atanasiu, N. (1962). *Phosphorsäure* **22**, 28–47; (1963). *C.A.* **58**, 2813.
18. Gericke, S. (1962). *Phosphorsäure* **22**, 48–60; (1963). *C.A.* **58**, 2813.
19. Askew, H. O. and Watson, J. (1946). *New Zealand J. Sci. Technol.* **28A**, 170–2.
20. Young, R. S. (1960). "Cobalt", Am. Chem. Soc. Monograph 149. Reinhold, New York.
21. Andrews, E. D. and Pritchard, A. M. (1947). *New Zealand J. Agric.* **75**, 501, 503–6.
22. Srinath, E. G. and Pillai, S. C. (1966). *Current Sci.* **35**, 247–50.
23. Andrezejewski, M., Czekalski, A., and Kocialkowski, Z. (1964). *Poznan, Towarz. Przyjaciol Nauk, Wydzial Nauk Roln. Lesnych, Prace Komisji Nauk Roln. Komisji Nauk Lesnych* **18**, 51–8; (1964). *C.A.* **61**, 8841.
24. Tokovoi, N. A., Maiboroda, N. M., and Lapshina, L. N. (1964). Mikroelem. Biosfere Ikh Primen. Sel'. Khoz. Med. Sib. Dal'nego Vostoka, Dokl. Sib. Konf. 2nd, 542–6; (1969). *C.A.* **70**, 86673.
25. Mikhailov, A. S. (1964). *Izv. Sib. Otdel. Akad. Nauk S.S.S.R., Ser. Biol.-Med. Nauk* No. 3, 101–4; (1965). *C.A.* **63**, 12273.
26. Khakimova, V. K., Vishnyakova, A. A., and Sultanov, S. (1972). *Uzbek. Khim. Zh.* **16**, 3–4; (1972). *C.A.* **77**, 18631.
27. Swaine, D. J. (1962). "The Trace-Element Content of Fertilizers", Commonw. Bur. Soil Sci. Tech. Commun. No. 52.
28. Gericke, S. (1956). Centre intern. Engrais chim. Assemblée gén. Rappt. 1, 185–93; (1959). *C.A.* **53**, 12547.
29. Cawse, P. A. (1977). Report AERE-R 8869. HMSO, London.

4 Cobalt in Waters

Traces of cobalt are found in sea waters, and all forms of fresh water and waste water. Table 3 gives some of the many concentrations of cobalt in waters reported over recent years. The content of cobalt in water, like that of other microelements, is usually expressed as $\mu g \, l^{-1}$.

Table 3 Cobalt Content of Waters

Water	Cobalt ($\mu g \, l^{-1}$)	Ref.
Sea water		
Oceanic	0·1	1
English Channel and Scottish coast	0·3	2
Coast of Japan	0·38–0·67	3
Puget Sound	0·23–0·32	4
Pacific, San Francisco	0·038	5
Black Sea	1·6–4·6	6
Azov Sea	2·4–4·5	6
Mediterranean Sea	0–3	7
Red Sea	0·40–1·50	8
Gulf of Aden	0·32–0·80	8
North-central Pacific	0·01	9
Tropical north-east Pacific	0·02	9
Oregon coast	0·046	9
Tropical north Atlantic	0·025	9
Bering Sea	0·030	9
Mediterranean Sea	0·020	9
Central Atlantic and Pacific	0·03–2·9	10
Fresh water		
Artesian wells, Kiev	0·48–2·22	11
Artesian wells, Leningrad	0·84–2·2	12
Drinking water, Transcarpathian Region	0·17	13
Drinking water, Lake Constance, purified	2·7	14
Drinking water, Kazakhstan	0–18	15
Ground water, Kharkov	0·16–0·78	16
Ground water, Zeravshan area	4·2	17
Irrigation water, Uzbekistan	0·1–10	18

Table 3　Cobalt Content of Waters (cont'd)

Water	Cobalt (μg l^{-1})	Ref.
Lake waters in Kazakhstan	0·41–9·1	19
Lake Constance, untreated	3·5	14
Mine water, Montana and Australia	4600–310,000	20
Mine water, Russia	9·7–100	21
Mineral water, Russia	400	22
Reservoir water, Russia	100	23
Reservoir water, Russia	4–30·9	24
Rivers, Voronezh Region	1·1–2·2	25
Rivers, Kazakhstan Region	6·4	19
Rivers, Leningrad Region	0·48–2·0	12
Rivers, Khar'kov Region	4–30·9	24
Rivers, Upper Danube	5·2	14
Rivers, North America	0–5·8	26
Rivers, U.S.A., France, Brazil	0·03–0·44	27
Springs, Virginia	200	20
Springs, Japan	1–6	28
Springs, Russia	1–9·8	21
Subsoil water, Upper Amur Region	tr.–3	29
Subsoil water, Kedabek Region	2·9–86	30
Surface water, general	2·1	31
Surface water, Leningrad Region	0·38–4·0	32
Underground water, Voronezh Region	0·8–2·3	33
Water samples, Azerbaidzham	2–10	34
Water samples, Azerbaidzham	1–4	35
Well shafts, Leningrad Region	0·4–0·9	12

It will be seen that the values for cobalt in sea water vary appreciably, from about 0·01–4·6 μg l^{-1}.[1-9] This is to be expected owing to the variations in land drainage, currents, rock composition of the shore and sea bed, pollution, phytoplankton and marine organisms, and other factors which can contribute to the heterogeneity of sea water samples from different parts of the world.

Several workers have reported an increase in cobalt content in sea water with depth,[6,7,9,36] which is perhaps associated with a decrease in phytoplankton, and an increase from central to coastal regions.[6,9] In the deep sea, much of the cobalt may be removed by co-precipitation with manganese oxides.

Data for fresh waters show an even greater range in cobalt content.[11-35] Most values fall within about 0·1–10 μg l^{-1}, but some spring, mineral, and mine waters contain much more cobalt. Some mineral waters, however, are very low in cobalt; about half of 126 samples of Bulgarian mineral waters failed to reveal cobalt when tested by spectrographic analysis.[36] It has been reported that acidic springs have a high cobalt concentration, and that in mine waters the amount of cobalt increases with the sulphate content.[21] Other workers have found that water from coal mines contained large amounts of cobalt, which was derived from the minerals occurring naturally in the coal beds.[37]

Several workers have observed that deep-water sources have a higher cobalt content than surface waters,[32,38,39] but it has also been reported that no correlation could be found between the cobalt content of underground waters and any definite stratigraphic horizon.[25]

There have been reports that waters from goitrogenic areas were low in cobalt, and that in regions where the water contained normal amounts of cobalt and boron, even when iodine was below the limit, goiter was not prevalent;[39] other workers, however, have been unable to find any correlation between the cobalt content of water, and the occurrence of goiter.[15]

The behavior of cobalt in reservoirs has been examined. Little change was found in cobalt throughout the year.[23] Another study indicated that, on average, 0.086% of cobalt is bound up with plankton, and about 12.9% of the element contained in the organisms enter as vitamin B_{12}; when the concentration of cobalt in the reservoir is raised artificially, its reserves concentrate in the mud.[40]

The distribution of cobalt between liquid and suspended particles in Wisconsin lakes has been described.[41] In the Amazon and Yukon Rivers, cobalt is distributed equally between precipitated metallic coatings and crystalline solids.[42]

Although the subject of manganiferous nodules found on the deep-sea floors is primarily one concerning mineral deposits and extractive metallurgy, and accordingly outside the field of this monograph, it may merit mention here. These nodules contain, among other metals, about $0.01–2.3\%$ cobalt; most of the latter is in the manganese oxide phases. It has been estimated that the reserves of cobalt in ocean nodules are growing at an annual rate of 31.5×10^6 kg; this is about double the present world consumption of cobalt. Many publications and patents attest to the current interest in this subject.[43–45]

The cobalt content of boiler feed water in the high pressure region has been reviewed.[46] Eolian dust added to ^{60}Co-tagged sea-water samples, in amounts of $0.005–1$ g per 50 ml, at temperatures of 8–20 °C, removed almost 56% of the cobalt present in 48 h; this was attributed to surface adsorption on the dust.[47]

References

1. Goldberg, E. D. (1954). *J. Geol.* **62**, 249–65.
2. Black, W. A. P. and Mitchell, R. L. (1952). *J. Marine Biol. Ass. U. K.* **30**, 575–84.

3. Ishibashi, M. (1953). *Records Oceanogr. Wks Japan* **1**, 88–93; (1954). *C.A.* **48**, 6175.
4. Thompson, T. G. and Laevastu, T. (1960). *J. Marine Res.* **18**, 189–92.
5. Weiss, H. V. and Reed, J. A. (1960). *J. Marine Res.* **18**, 185–8.
6. Rozhanskaya, L. I. (1963). *Trudȳ Sevastopol'sk Biol. St., Akad. Nauk S.S.S.R.* **16**, 467–71; (1964). *C.A.* **61**, 8056.
7. Rozhanskaya, L. I. (1965). Osnovye Cherty Geol. Stroeniya, Gidrolog. Rezhima i Biol. Sredizemn. Morya, Akad. Nauk S.S.S.R., Okeanogr. Komis., 146–9; (1966). *C.A.* **64**, 3202.
8. Rozhanskaya, L. I. (1966). *Gidrobiol. Zh., Akad. Nauk Ukr. S.S.R.* **2**, 40–2; (1966). *C.A.* **65**, 5214.
9. Robertson, D. E. (1970). *Geochim. Cosmochim. Acta* **34**, 553–67.
10. Schutz, D. F. and Turekian, K. K. (1965). *Geochim. Cosmochim. Acta* **29**, 259–313.
11. Barannik, P. I., Mikhalyuk, I. A., Mnatsakanyan, R. P., Tsvetkova, I. N., and Yatsula, G. S. (1961). *Gig. Sanit.*, 95–6; (1961). *C.A.* **55**, 21427.
12. Batmanova, O. Ya. (1965). *Trudȳ Leningrad. Sanit.-Gig. Med. Inst.* **81**, 46–9; (1967). *C.A.* **66**, 94272.
13. Meshchenko, V. M., Aleksik, V. I., and Mezhvinskaya, E. A. (1959). *Gig. Sanit.* **24**, 7–10; (1959). *C.A.* **53**, 17380.
14. Quentin, K. E. and Winkler, H. A. (1974). *Zentbl. Bakt. ParasitKde, Infektionskr. Hyg. Abt. 1: Orig., Reihe B* **158**, 514–23; (1974). *C.A.* **81**, 41206.
15. Chigrina, T. A., Makushinskaya, N. D., and Shumaev, V. D. (1974). *Trudȳ Nauch.-Issled. Inst. Kraev. Patol., Alma-Ata* **26**, 66–9; (1976). *C.A.* **84**, 169383.
16. Litvinenko, M. G. (1958). Materialy Nauch. Konf. Sanit.-Gig. Fakul'teta, Posvyaschen. 40-Letiyu Velikoi Oktyabr. Sots. Revolyutsii, Kharkov Med. Inst., Sbornik, 8–9; (1960). *C.A.* **54**, 25404.
17. Kozyreva, G. F. (1965). Mikroelementy v Sel'. Khoz. Akad. Nauk Uzbet. S.S.R., Otdel. Khim.-Tekhnol. i Biol. Nauk, 313–17; (1966). *C.A.* **64**, 13335.
18. Mavlyanov, G. A. and Mirzaeva, K. Kh. (1963). *Dokl. Akad. Nauk Uzbek. S.S.R.* **20**, 40–2; (1963). *C.A.* **59**, 13687.
19. Idrisova, R. A., Bekturov, A. B., and Mun, A. I. (1964). *Trudȳ Inst. Khim. Nauk, Akad. Nauk Kaz. S.S.R.* **10**, 88–91; (1964). *C.A.* **61**, 11750.
20. Clarke, F. W. (1916). "The Data of Geochemistry", U.S. Geol. Surv. Bull. 616.
21. Vekilova, F. I., Borovskaya, Y. B., and Efendieva, E. K. (1962). *Izv. Akad. Nauk Azerb. S.S.R., Ser. Geol.-Geogr. Nauk Nefti* No. 2, 43–52; (1962). *C.A.* **57**, 12260.
22. Raspopov, E. I. (1961). *Trudȳ Novocherk. Politekhn. Inst.* **21**, 29–34; (1962). *C.A.* **58**, 344.
23. Lifshits, G. M. (1961). *Gidrokhim. Mater.* **33**, 75–9; (1962). *C.A.* **57**, 8360.
24. Litvinenko, M. G. (1969). *Gig. Naselennykh Mest* **8**, 5–8; (1970). *C.A.* **73**, 48400.
25. Lifshits, G. M. (1961). *Trudȳ Voronezh. Zoovet. Inst.* No. 1, 111–17; (1963). *C.A.* **58**, 344.
26. Durum, W. H. and Haffty, J. (1963). *Geochim. Cosmochim. Acta* **27**, 1–11.
27. Kharkar, D. P., Turekian, K. K., and Bertine, K. K. (1968). *Geochim. Cosmochim. Acta* **32**, 285–98.
28. Torii, T. (1955). *J. Chem. Soc. Japan, Pure Chem. Sect.* **76**, 707–10; (1956). *C.A.* **50**, 11569.
29. Pryakhin, A. I., Chekhovskikh, M. M., and Shchebunyaeva, I. A. (1963). *Izv. Vȳssh. Ucheb. Zaved., Geol. i Razvedka* **6**, 90–8; (1963). *C.A.* **59**, 1374.
30. Malyuga, D. P. and Niyazov, A. Kh. (1971). *Izv. Akad. Nauk Azerb. S.S.R., Ser. Biol. Nauk* **2**, 72–8; (1972). *C.A.* **76**, 13127.

31. Malyuga, D. P. (1949). *Akad. Nauk S.S.S.R.* **67**, 1057–60; (1950). *C.A.* **44**, 496.
32. Batmanova, O. Ya. (1963). Materialy Resp. Itog. Nauchn. Konf. po Gigiene, Leningrad Sb., 27–8; (1964). *C.A.* **61**, 8053.
33. Lifshits, G. M. (1961). *Trudȳ Voronezh. Zoovet. Inst.* **17**, 101–9; (1963). *C.A.* **58**, 344.
34. Gyul'akhmedov, A. N. and Gadzhiev, F. M. (1968). Mater. Respub. Konf. Probl. "Mikroelem. Med. Zhivotnovod." 1st, 11–13; (1970). *C.A.* **73**, 55056.
35. Dzhafarov, Ya. M. and Gyul'akhmedov, A. N. (1968). Mater. Respub. Konf. Probl. "Mikroelem. Med. Zhivotnovod." 1st, 13–14; (1970). *C.A.* **73**, 55057.
36. Penchev, N. P., Pencheva, E. N., and Bonchev, P. R. (1960). *C.R. Acad. Bulg. Sci.* **13**, 55–7; (1961). *C.A.* **55**, 10758.
37. Klimov, I. T. and Fesenko, N. G. (1961). *Gig. Sanit.* **26**, 97–8; (1961). *C.A.* **55**, 20272.
38. Rozhanskaya, L. I. (1967). Gidrofiz. Gidrokhim. Issled. Chernom More, 60–2; (1968). *C.A.* **69**, 89618.
39. Ancusa, M. and Pirvu, F. (1963). *Acad. Rep. Populare Romine, Baza Cercetari Stint. Timisoara, Studii Cercetari Biol. Stinte Agr.* **10**, 267–80; (1964). *C.A.* **60**, 15587.
40. Koval'skii, V. V. and Letunova, S. V. (1966). *Geokhimiya* **12**, 1478–88; (1967). *C.A.* **66**, 98350.
41. Parker, M. (1966). "Distribution of cobalt in lakes", U.S. At. Energy Commission; (1968). *C.A.* **69**, 6960.
42. Gibbs, R. J. (1973). *Science* **180**, 71–3.
43. Mero, J. L. (1965). "The Mineral Resources of the Sea". Elsevier, New York.
44. Agarwal, J. C., Beecher, N., Davies, D. S., Hubred, G. L., Kakaria, V. K., and Kust, R. N. (1976). *J. Metals* **28**, 24–31.
45. Sridhar, R., Jones, W. E., and Warner, J. S. (1976). *J. Metals* **28**, 32–7.
46. Stanisavlievici, L. (1959). *Energie* **11**, 581–2; (1960). *C.A.* **54**, 18837.
47. Aston, S., Chester, R., and Johnson, L. R. (1972). *Nature, Lond.* **235**, 380–1.

5 Effect of Cobalt on Microorganisms

Numerous interesting relationships between cobalt and various microorganisms have been reported. In this chapter will be summarized, in turn, the effect of cobalt on actinomycetes, algae, bacteria, fungi, and yeasts.

Actinomycetes

A strain of actinomycetes has been reported to tolerate as much as 10% cobalt chloride in a standard medium.[1] Cultures of actinomycetes from cobalt-containing muds in stagnant reservoirs have been found to produce vitamin B_{12}.[2]

Addition of $0 \cdot 05$–5 parts 10^{-6} cobalt, as cobalt chloride, to a culture medium rapidly increased vitamin B_{12} synthesis from actinomycetes, but propagation was not augmented. Above 5 parts 10^{-6}, both synthesis and propagation were inhibited in Czapek medium, but not in bouillon–peptone medium at concentrations below 20 parts 10^{-6}.[3]

The addition of $0 \cdot 25$–40 parts 10^{-6} cobalt to actinomycetes in a liquid medium partially or completely inhibited the biosynthesis of lytic enzymes, as well as that of protease and glucanase.[4]

Algae

An early study on the effect of cobalt on algae showed that this element stimulated growth in three species at concentrations of $0 \cdot 002$–$0 \cdot 2$ parts 10^{-6} cobalt, depending on the species; the addition of 2 parts 10^{-6}, or more, inhibited growth.[5] It has been reported that $0 \cdot 4$ parts 10^{-6} cobalt is necessary for optimum growth of four algal species.[6]

The cobalt content of various marine algae was found to range from $0 \cdot 08$–$0 \cdot 28$ parts 10^{-6} in living matter, or $1 \cdot 7$–$5 \cdot 3$ parts 10^{-6} in the ash.[7] Marine algae have a relatively high cobalt content, the ratio Co : Ni approaching 1; in sea water it is about $0 \cdot 3$.

One publication states that a concentration of cobalt exceeding 0·04 parts 10^{-6} retarded the growth of the algae *Chlorella vulgaris*, *Pediastrum tetias*, and *Euglena viridis*.[8] Another paper reported that *Anabaena flos-aquae A-37* required less than 50 parts 10^{-9} of cobalt.[9] The re-establishment of algal growth in complete nutritive solutions containing a high concentration of cobalt indicates the important accumulation capacity of *Scenedesmus acutiformis* cells for this element.[10]

Bacteria

An early reference book on the chemistry and physiology of bacteria stated that the toxicity of cobalt towards bacteria varies with the species, and in general occupies an intermediate position among cations.[11] The same statement is broadly true today.

Many publications have reported that small concentrations of cobalt were beneficial, or at least not harmful, to bacteria. Growth of *Streptomyces griseus* and *B. subtilis* was not affected by 2 parts 10^{-6} cobalt.[12] Studies with *B. natto* showed cobalt was necessary for cell multiplication.[13] When cobalt was added to *B. asterosporus* in various sugar media, both acid production and sugar consumption were increased, and respiration coefficients were lowered.[14] Addition of cobalt to the culture medium for *Streptococcus faecalis* increased the level of acid pyrophosphatase activity.[15] The growth and porphyrin production of *Mycobacterium avium* were substantially increased by 0·27 parts 10^{-6} cobalt.[16] It was suggested that the effect of cobalt on *Brucella abortus* was an inhibition of the synthesis of a cyanide-sensitive respiratory enzyme.[17] The amount of cobaltous ion fixed by *E. coli* increased as the pH rose from 5·6–7·7, and was directly related to the cobalt concentration in the medium.[18] A mutant of *E. coli* has been found to be resistant to 5×10^{-4} M cobalt; the mechanism of cobalt-resistance did not appear to be the result of a cobalt-permeability barrier.[19]

When 2–120 mg cobalt l^{-1} were added to *Streptomyces olivaceus* and *B. megaterium*, the organisms adapted, but vitamin B_{12} production decreased.[20] Growth of *Propionibacterium freudenreichii* and its synthesis of vitamin B_{12} were stimulated by up to 3 mg l^{-1} of cobalt, but higher concentrations caused a decrease in both.[21] Cobalt added to *B. Calmette-Guerin* cultures gave an uptake of 0·68 parts 10^{-6} cobalt, and the microbial requirement for this element was placed at 0·5–1·5 parts 10^{-6} in the cells.[22] In the animal rumen, 30 mg cobalt chloride produced a 41 % increase in the total number of aerobic microorganisms, a 95 % increase of aerobic lactic acid bacteria, a 50% increase of total anaerobic organisms, and an 86 % increase of anaerobic lactic acid bacteria.[23] Cobalt is effective in promoting enzyme activity of

Pseudomonas aeruginosa mainly through binding rather than catalysis.[24] For a culture of *B. mesentericus*, cobalt was the most effective ion for stimulating the synthesis of protease.[25]

Some studies, however, have failed to show any beneficial effect of cobalt on bacteria. A concentration of 2×10^{-4} M cobalt caused the cytochrome in *Proteus vulgaris* to disappear, along with a reduction of catalase activity.[26] Oxygen uptake by the resting cells of *Shigella flexneri* was inhibited by 10^{-4} M cobalt.[27] The growth and viability of *B. Calmette-Guerin* was depressed by 1–2 mg cobalt chloride l^{-1}.[28] The inhibiting effect of L-serine and D-serine on the growth of *Staphylococcus aureus* is enhanced by cobalt, but this enhancement is completely suppressed by L-cysteine or L-histidine; this amino acid antagonism may be due to metal binding.[29] Cobalt seriously inhibits development of *E. coli*.[30] Zinc has an adverse effect on the growth of propionic acid bacteria, and the addition of cobalt does not diminish this effect.[31] Exposure to dilute solutions of cobalt sulphate decreased the growth and virulence of both typhoid and dysentery bacilli.[32] It has been reported that cobalt was required for cobalamin production, but not for the growth of *B. megatherium*.[33]

Nitrogen-fixing bacteria in nodules of legumes

A large proportion of the literature on the role of cobalt in bacterial growth and development is concerned with the activities of nitrogen-fixing bacteria on the root nodules of legumes. Virtually all papers report a beneficial effect from the addition of small quantities of cobalt. For soybeans, in the absence of supplied nitrogen, plants receiving 1–50 ng/g cobalt chloride showed normal vigorous growth, whereas a lower content resulted in poor growth and nitrogen deficiency; lack of cobalt seems to affect the symbiosis between the legume and the nodule bacteria.[34] Addition of 0·006–0·06 parts 10^{-6} cobalt stimulated nitrogen-fixation of legumes.[35] The size of root nodules, and their hemoglobin content in lupine, are increased by 5×10^{-6} M cobalt in nutrient solutions for sand cultures.[36] Addition of 0·5 mg cobalt kg^{-1} soil improved the number and appearance of nodules, and increased the yield of alfalfa.[37] In extracts of *Clostridium tetanomorphum*, cobalt porphyrin synthase inserts Co^{2+} into many dicarboxylic porphyrins.[38] A variety of different nitrogen-fixing organisms, such as *Azotobacter*, *Clostridium*, and *Rhizobium*, require cobalt for growth and for the synthesis of vitamin B_{12} compounds.[39]

Nitrogen fixation by *Azotobacter* was stimulated by cobalt; the element was found in the ash of cells grown in media containing cobalt.[40] Cobalt was shown to be a requirement for the growth of several bacteria grown on nitrate, and less cobalt was required for *Azotobacter vinelandii* grown on NH_4^+ and glutamate; nitrate reductase formation depended on cobalt.[41] Favorable effects of cobalt on both the development of *Azotobacter* and its nitrogen-fixing activity have been demonstrated.[42] Nitrogen fixation by *Azotobacter chroococcum* was enhanced 50 % by 0·1 part 10^{-6} cobalt, but somewhat less by 1–5 parts 10^{-6}.[43] Another researcher working with *Azotobacter chroococcum* found that 0·1–1 parts 10^{-6} cobalt increased cell growth and nitrogen fixation, but 5 parts 10^{-6} depressed them.[44]

Cobalt also increased nitrogen fixation on legumes by *Rhizobium* species. The omission of cobalt resulted in nitrogen starvation, and a minute addition increased weight and nitrogen in plant and nodules by 100 % in 7 weeks.[45] Total nitrogen content of the cells from cultures treated with 0·5–5 parts cobalt 10^{-9} parts medium was about ten times greater than that of cells from cultures lacking cobalt.[46] In a study of the root nodules of *Alnus*, a sand culture of *Alnus glutinosa* without cobalt gave a weight of 3·54 g with 32·7 mg nitrogen, whereas with 0·01 part 10^{-6} cobalt the weight was 8·60 g, and the nitrogen 139·0 mg.[47] The addition of 0·1 part 10^{-9} cobalt to *Rhizobium meliloti* and other species caused a marked growth increase; no other metal could be substituted successfully.[48]

A deficiency of cobalt markedly reduced nitrate reductase activity in *Rhizobium*.[49] In *Rhizobium meliloti*, 0·02–2 parts 10^{-9} cobalt increased bacterial growth, but 0·2 parts 10^{-6} was inhibitory; the vitamin B_{12} content of cells markedly increased with increasing cobalt content.[50] Addition of cobalt to legumes resulted in a 40 % increase in dry matter yield and an increase in protein; all strains of *Rhizobium* accumulated cobalt in the nodule.[51] One paper concluded that cobalt deficiency in *Rhizobium meliloti* prevents the synthesis of quantities of vitamin B_{12} coenzyme adequate for the normal function of methylmalonyl coenzyme A mutase, and that the inactive mutase results in the failure of the organism to oxidize propionate.[52] In lupine, *Rhizobium* and *Azotobacter* gave increased dehydrogenase activity when cobalt was added at 0·001–0·01 mg l^{-1}.[53] For *Rhizobium*, maximum biomass and oxidase activity was given by 0·01–0·1 mg cobalt l^{-1}; dehydrogenase activity was increased 40 % by 0·01 mg cobalt l^{-1} but 0·1 mg l^{-1} caused inhibition.[54]

Cobalt enlarges root nodules, and affects growth and development of yellow lupine; it brings about regressive changes in the stimulants of phytopathological processes in plant tissues.[55] Cobalt stimulates the nitrate reductase activity in leaves and nodules of legumes; the application of toxic amounts causes chlorosis and a decrease in enzyme activity.[56] Nodulation and nitrogen-fixation in peas were both improved by adding 30 μg cobalt per 3·5 kg sand.[57] Under aseptic conditions, the highest yield and nitrogen content of soybeans were found when cobalt was added at the rate of 20–400 mg kg^{-1} of seeds.[58]

Cobalt was found to have a beneficial effect on the growth of *Trifolium subterraneum*, whether or not the plants are nodulated, and this effect is largely independent of the effectiveness of the rhizobial symbiont used.[59] A study on the distribution of cobalt in *Trifolium subterraneum* showed that nodules, and, to a lesser extent, the roots, accumulated much more cobalt than did the leaves.[60] Nitrogen accumulation in legumes was increased by over 100% after addition of cobalt.[61] *Mycobacterium flavum* cultured in a nitrogen-free medium with pyruvate did not react to the addition of cobalt, but when cultured in a medium containing molybdenum and ethanol, the addition of cobalt increased nitrogen fixation by 650%.[62] In non-nodulating *Trifolium subterraneum* grown with nitrate or urea nitrogen, a cobalt deficiency was produced; deficiency effects were also obtained in sterile cultures of this clover utilizing either cobalt sulphate or cyanocobalamin as a source of cobalt.[63] Cobalt increased the size of root nodules in peas, and the size and number of nodules in beans; the element also reversed the inhibiting effect of high nitrogen concentrations on nodule formation.[64]

Several investigators found that cobalt had an effect on the action of antibiotics. A 2- to 15-fold increase in the bacteriostatic action of several antibiotics on *E. coli* and *Micrococcus pyogenes aureus* has been reported.[65] A similar action has been observed against various bacteria with cobalt additions to penicillin, streptomycin, and bacitracin.[66] Cobalt increases the inhibitory effect of penicillin against *Salmonella pullorum*. Concentrations of 0·125–0·25 parts 10^{-6} cobalt have been observed to increase the effect of penicillin on *Staphylococcus aureus*, *Streptococcus pyogenes*, *E. coli*, and *Klebsiella* sp.[68] It was found that cobalt did not accelerate the development of resistance to streptomycin by various bacteria.[69] In the presence of cobalt chloride, the sensitivity of *Staphylococcus aureus* towards penicillin increased.[70]

On the other hand, one worker reported that 0·005% cobalt chloride reduced the activity of penicillin by a factor of 2–4; lower concentrations did not alter the activity. Concentrations of cobalt up to 0·005% did not reduce the activity of streptomycin with respect to hemolytic streptococci, nor the activity of oxytetracycline on staphylococci and hemolytic streptococci.[71]

Fungi

The effect of small quantities of cobalt on various fungi has interested a number of investigators. Early work indicated that increments of cobalt sulphate up to 0·002% increased the growth and weight of *Aspergillus niger* and *Penicillium glaucum*, but a concentration of 0·033% reduced growth to below that of the check.[72] Another early worker found that the yield of *Aspergillus niger* decreased consistently with increasing additions of 0·1–

50 mg cobalt nitrate l^{-1} of nutrient solution.[73] The toxic limits of cobalt for *Aspergillus niger*, *Penicillium oxalicum*, and *P. expansum* have been given as 1500–1600 parts 10^{-6}; cobalt salts were less injurious to *Penicillium* than were either mercury or silver, but more detrimental than either cadmium, lead, or nickel.[74] One paper reported that the presence of ferrous sulphate had a depressing effect on the yield of ascorbic acid in *Aspergillus flavus*, but this effect was more than overcome by adding cobalt to the medium.[75]

Cobalt toxicity reduced the catalase activity of *Neurospora crassa* to 12, 18, and 70% of control values with NO_3, $NO_3 + NH_4$, and NH_4 media, respectively.[76] A slow uptake *in vivo* of cobalt from a growth medium resulted in an increased density of mitochondria of *Schizophyllum commune*.[77]

The growth of mycelium in *Fusarium blasticola* was slowed by 75% with 0·7 mg cobalt chloride ml^{-1} of liquid medium.[78] The development of cotton wilt was slower, and its manifestations markedly milder in plants grown with 1–2 mg cobalt kg^{-1} of sand. A solution of 0·1% cobalt retarded the growth of the fungus *Sclerotinia*, responsible for storage rot of sunflower.[80]

Yeasts

A few workers have studied the role of cobalt in the growth and development of yeasts. Only 10% of the cobalt assimilated by the yeast *Saccharomyces cerevisiae* was adsorbed on the surface, the rest being combined chemically.[81] It has been reported that treatment of yeasts with a small quantity of cobalt increases cell numbers, cell sizes, and proteins in the resulting yeast-containing products.[82] It has also been observed that nitrogen is higher in cobalt-containing yeast than in the normal product.[83] *Saccharomyces cerevisiae* grown in media containing progressively higher concentrations of inorganic cobalt up to 750 parts 10^{-6}, finally attained a content of 9·9% cobalt in the yeast cells.[84]

Cobalt enhances the production of riboflavin, and the uptake of iron and magnesium by the yeast *Candida guilliermondii*.[85] Cobalt chloride brings about a respiratory deficiency in yeast.[86] The addition of cobalt to a nutrient solution of yeast increased the biosynthesis of zymosan.[87] The ash of dry yeast-extract contained 3·5 parts 10^{-6} cobalt.[88] The presence of cobalt in the culture liquid of *Saccharomyces cerevisiae* increased the immunobiological activity of zymosan produced by the cells.[89] Small additions of cobalt stimulated the growth of *Saccharomyces cerevisiae* and the biosynthesis of vitamin B_{12}; larger additions had an inhibitory effect.[90] Another worker reported that 0·0002–10 mg cobalt l^{-1} resulted in a 16–40% increase in the biomass of *Saccharomyces cerevisiae*.[91]

Reviews

Reviews have appeared on the influence of cobalt on microorganisms and their adaptability to the natural concentrations of cobalt in a medium,[92] and on cobalt in the nutrition of microorganisms.[93] A correlation has been found between cobalt in the humus layer and the microflora of soil aggregates.[94] It has been suggested that the protective effect of cobalt can account for the increased production of riboflavine by some organisms in the presence of an optimal concentration of Co^{2+}.[95]

References

1. Kojima, H. and Matsuki, M. (1956). *Tohoku J. Agric. Res.* 7, 175–87; (1957). *C.A.* 51, 10642.
2. Letunova, S. V. (1958). *Mikrobiologiya* 27, 429–34; (1958). *C.A.* 52, 18660.
3. Uesaka, S., Kawashima, R., and Hashimoto, Y. (1958). *Kyoto Daigaku Shokuryo Kagaku Kenkyusho Hokoku* No. 21, 1–7; (1959). *C.A.* 53, 5436.
4. Shukan, L. A. and Shklyar, B. Kh. (1975). Biol. Akt. Veshchestva Mikroorg., 44–9; (1976). *C.A.* 85, 119287.
5. Young, R. S. (1935). Cornell Univ. Agric. Exp. *Stn Mem.* 174.
6. Holm-Hansen, O., Gerloff, G. C., and Skoog, F. (1954). *Physiologia Pl.* 7, 665–75; (1956). *C.A.* 50, 1138.
7. Vinogradov, A. P. (1953). "The Elementary Chemical Composition of Marine Organisms". Sears Foundation for Marine Research, New Haven.
8. Coleman, R. D., Coleman, R. L., and Rice, E. L. (1971). *Bot. Gaz.* 132, 102–9.
9. Tiranasar, P. C. and Tischer, R. G. (1973). *J. Miss. Acad. Sci.* 19, 107–17.
10. Peterfi, S., Nagy-Toth, F., and Barna, A. (1975). *Studia Univ. Victor Babes, Bolyai, Ser. Biol.* 20, 17–23; (1976). *C.A.* 84, 42496.
11. Porter, J. R. (1946). "Bacterial Chemistry and Physiology". Wiley, New York.
12. Tanaka, S., Sawada, Y., Nozaki, Y., and Yamamoto, T. (1954). *J. Chem. Soc. Japan, Pure Chem. Sect.* 75, 252–4; (1954). *C.A.* 48, 12899.
13. Sawada, Y., Tanaka, K., Hirano, M., Sato, M., Miyamoto, J., and Tanaka, S. (1955). *J. Chem. Soc. Japan, Pure Chem. Sect.* 76, 274–7; (1957). *C.A.* 51, 18107.
14. Dedic, G. A. and Koch, O. G. (1955). *Arch. Mikrobiol.* 23, 130–41; (1958). *C.A.* 52, 18648.
15. Oginsky, E. L. and Rumbaugh, H. L. (1955). *J. Bact.* 70, 92–8.
16. Patterson, D. S. P. (1959). *Nature, Lond.* 185, 57.
17. Altenbern, R. A., Williams, D. R., and Ginoza, H. S. (1959). *J. Bact.* 77, 509.
18. Katayama, Y. (1960). *Japan J. Microbiol.* 4, 277–82; (1962). *C.A.* 56, 2762.
19. Katayama, Y. (1960). *Japan J. Microbiol.* 4, 351–6; (1962). *C.A.* 56, 5202.
20. Koval'skii, V. V. and Letunova, S. V. (1963). *Mikrobiologiya* 32, 850–5; (1964); *C.A.* 61, 977.
21. Rao, S. and Washington, D. R. (1964). *Nature, Lond.* 202, 212–13.
22. Sternberg, J., Benoit, J. C., Mercier, A., and Paquette, J. C. (1964). *Rev. Can. Biol.* 23, 353–65.
23. Svegzdaite-Laurinavichine, D. (1965). *Liet. TSR Mokslų Akad. Darb., Ser. C* No. 2, 3–15; (1966). *C.A.* 64, 13121.

24. Morihara, K. and Tsuzuki, H. (1974). *Agric. Biol. Chem.* **38**, 621–6; (1974). *C.A.* **81**, 22552.
25. Ciurlys, T., Uzkurenas, A., and Povilaitiene, J. (1975). *Liet. TSR Mokslų Akad. Darb.*, *Ser. C* No. 3, 135–46; (1975). *C.A.* **83**, 160425.
26. Petras, E. (1957). *Arch. Mikrobiol.* **28**, 138–44; (1959). *C.A.* **53**, 16273.
27. Kitamura, M.(1959). *J. Okayama Med. Soc.* **71**, 3671–80; (1960). *C.A.* **54**, 25054.
28. Goncharevskaya, T. S., Salivon, E. S., Slyusarenko, I. T., Gorodetskaya, P. M., and Evalenko, E. S. (1961). *Zh. Mikrobiol. Epidem. Immunobiol.* **32**, 70–5; (1961). *C.A.* **55**, 26117.
29. Weinberg, E. D. (1960). *Antonie van Leeuwenhoek. J. Microbiol. Serol.* **26**, 321–8; (1961). *C.A.* **55**, 9551.
30. Fauget, M. and Goudot, A. (1961). *Annls Inst. Pasteur, Paris* **101**, 862–8; (1962). *C.A.* **56**, 14733.
31. Vorob'eva, L. I. (1962). *Dokl. Akad. Nauk S.S.S.R.* **145**, 1381–4; (1962). *C.A.* **57**, 17179.
32. Priselkov, M. M. and Grigor'eva, V. M. (1955). *Zh. Mikrobiol. Epidem. Immunobiol.* No. 3, 70–6; (1955). *C.A.* **49**, 14902.
33. Garibaldi, J. A., Ijichi, K., Snell, N. S., and Lewis, J. C. (1953). *Ind. Engng Chem.* **45**, 838–46.
34. Ahmed, S. and Evans, H. J. (1959). *Biochem. Biophys. Res. Commun.* **1**, 271–5.
35. Hallsworth, E. H., Wilson, S. B., and Greenwood, E. A. N. (1960). *Nature, Lond.* **187**, 79–80.
36. Michiels, L. and Sironval, C. (1960). *Archs Intern. Physiol. Biochim.* **68**, 843–4; (1961). *C.A.* **55**, 9587.
37. Melkumova, T. M. and Gazaichyan, Zh. M. (1964). *Dokl. Akad. Nauk Azerb. S.S.R.* **20**, 53–7; (1964). *C.A.* **61**, 8847.
38. Porra, R. J. and Ross, B. D. (1965). *Biochem. J.* **94**, 557–62.
39. Evans, H. J., Russell, S. A., and Johnson, G. V. (1965). Non-Heme Iron Proteins, Role Energy Conversion Symp., Yellow Springs, Ohio, 303–13; (1966). *C.A.* **64**, 5484.
40. Iswaran, V. and Sundara Rao, W. V. B. (1960). *Proc. Indian Acad. Sci., Part B* **51**, 103–15.
41. Nicholas, D. J. D., Kobayashi, M., and Wilson, P. W. (1962). *Proc. Nat. Acad. Sci. U.S.* **48**, 1537–42.
42. Abugalybov, M. (1961). *Trudy Nauch.-Issled. Inst. Zemledeliya, Azerb. Akad. Sel'.-khoz. Nauk* **6**, 3–13; (1963). *C.A.* **58**, 9433.
43. Iswaran, V. and Sundara Rao, W. V. B. (1964). *Nature, Lond.* **203**, 549.
44. Kurdina, R. M. (1968). *Trudy Inst. Mikrobiol. Virus., Akad. Nauk Kazak. S.S.R.* **11**, 84–6; (1968). *C.A.* **69**, 50151.
45. Reisenauer, H. M. (1960). *Nature, Lond.* **186**, 375–6.
46. Lowe, R. H., Evans, H. J., and Ahmed, S. (1960). *Biochem. Biophys. Res. Commun.* **3**, 675–8.
47. Bond, G. and Hewitt, E. J. (1962). *Nature, Lond.* **195**, 94–5.
48. Lowe, R. H. and Evans, H. J. (1962). *J. Bact.* **83**, 210–11.
49. Nicholas, D. J. D., Maruyama, Y., and Fisher, D. J. (1962). *Biochim. Biophys. Acta* **56**, 623–6.
50. Kliewer, W. M. and Evans, H. J. (1962). *Archs Biochem. Biophys.* **97**, 427–9; (1962). *C.A.* **57**, 5107.
51. Hallsworth, E. G. and Wilson, S. B. (1962). *Proc. Univ. Nottingham Easter School Agric. Sci.* **9**, 44–6; (1964). *C.A.* **61**, 6055.

52. De Hertogh, A. A., Mayeux, P. A., and Evans, H. J. (1964). *J. Biol. Chem.* **239**, 2446–53.
53. Koleshko, O. I. (1970). *Fiziol. Biokhim. Mikroorg.*, 177–81; (1972). *C.A.* **76**, 1642.
54. Koleshko, O. I. and Danil'chik, N. I. (1966). Mikroelem. Sel'. Khoz. Med., Dokl. Vses. Soveshch. Mikroelem. 5th, 474–8; (1970). *C.A.* **72**, 129613.
55. Yaroshenko, T. V. (1963). *Zashch. Rast. Vredit. Bolez.* **8**, 22–4; (1965). *C.A.* **62**, 11105.
56. Yagodin, B. A., Ovcharenko, G. A., Vasil'eva, Yu. Y., and Ivanova, M. A. (1970). *Sel'.-khoz. Biol.* **5**, 134–6; (1970). *C.A.* **73**, 34329.
57. Vanek, V. and Knop, K. (1972). *Rostl. Výroba* **18**, 521–9; (1972). *C.A.* **77**, 163531.
58. Chomchalow, S. (1975). *Thai J. Agric. Sci.* **8**, 1–5; (1975). *C.A.* **83**, 112845.
59. Wilson, S. B. and Hallsworth, E. G. (1965). *Pl. Soil* **22**, 260–79; (1967). *C.A.* **67**, 107736.
60. Wilson, S. B. and Hallsworth, E. G. (1966). *Pl. Soil* **23**, 60–78; (1967). *C.A.* **67**, 88318.
61. Danilova, T. A. and Demkina, E. N. (1967). *Dokl. Akad. Nauk S.S.S.R.* **172**, 487–90; (1967). *C.A.* **66**, 75320.
62. Il'ina, T. K. (1967). *Mikrobiologiya* **36**, 626–31; (1967). *C.A.* **67**, 106159.
63. Wilson, S. B. and Nicholas, D. J. D. (1967). *Phytochemistry* **6**, 1057–66.
64. Sadovskaya, E. N. and Peterburgskii, A. V. (1975). *Dokl. TSKHA* **213**, 36–42; (1976). *C.A.* **85**, 4336.
65. Forni, P. V. (1953). *Boll. Soc. Ital. Patol.* **3**, 183–4; (1956). *C.A.* **50**, 12172.
66. Trace, J. C. and Edds, G. T. (1954). *Am. J. Vet. Res.* **15**, 639–42.
67. Pital, A., Stafseth, H. J., and Lucas, E. H. (1953). *Science* **117**, 459–60.
68. Vargas, P. N., Munoz, A., and de Moreno, J. C. (1958). *Archs Int. Pharmacodyn. Thér.* **116**, 1–16; (1959). *C.A.* **53**, 4428.
69. Chernomordik, A. B. and Kobeleva, P. S. (1959). *Antibiotiki* **4**, 96–8; (1960). *C.A.* **54**, 19845.
70. Mnatsakanov, S. T. (1967). *Antibiotiki* **12**, 161–2; (1967). *C.A.* **66**, 83317.
71. Vasil'eva, N. V. (1965). Antibiotiki, Akad. Nauk Ukr. S.S.R., Inst. Mikrobiol., 115–18; (1966). *C.A.* **64**, 7075.
72. Beeson, K. C. (1950). U.S. Dept. Agric. Inf. Bull. 7.
73. Steinberg, R. A. (1920). *Bot. Gaz.* **70**, 465–8.
74. Bedford, C. L. (1936). *Zentbl. Bakt. ParasitKde, Abt. II* **94**, 102–12.
75. Ramakrishnan, C. V. and Desai, P. J. (1956). *Current Sci.* **25**, 189–90.
76. Subramanian, K. N. and Sarma, P. S. (1968). *Biochim. Biophys. Acta* **156**, 199–202.
77. Watrud, L. S. and Ellingboe, A. H. (1973). *J. Cell Biol.* **59**, 127–33; (1973). *C.A.* **79**, 144255.
78. Uscuplic, M. (1961). *Zašt. Bilja* Nos 65–66, 113–21; (1963). *C.A.* **59**, 5533.
79. Poletaeva. V. F. (1969). *Izv. Akad. Nauk Turkmen. S.S.R., Ser. Biol. Nauk* **3**, 73–4; (1969). *C.A.* **71**, 109881.
80. Polyakov, P. V. (1971). *Khim. Sel'. Khoz.* **9**, 109–11; (1971). *C.A.* **74**, 124265.
81. Bass, H. and Zizuma, A. (1956). *Latv. PSR Zināt. Akad. Vest.* No. 8, 109–14; (1957). *C.A.* **51**, 9781.
82. Korshakov, P. N. (1953). *Zhivotnovodstvo* No. 5, 61–3; (1955). *C.A.* **50**, 505.
83. Erkama, J. and Enari, T. M. (1956). *Suom. Kemistibehti* **29B**, 176–8; (1957). *C.A.* **51**, 4494.
84. Perlman, D. and O'Brien, E. (1954). *J. Bact.* **68**, 167–70.

85. Enari, T. M. (1958). *Annls Acad. Sci. Fennicae, Sec. A II* No. 90, 8–42; (1959). *C.A.* **53**, 2361.
86. Lindgren, C. C., Nagai, S., and Nagei, H. (1958). *Nature, Lond.* **182**, 446–8.
87. Bass-Shadkhan, Kh. F. and Galina, I. (1961). *Latv. PSR Zināt. Akad. Vest.* No. 12, 69–74; (1962). *C.A.* **56**, 14729.
88. Grant, C. L. and Pramer, D. (1962). *J. Bact.* **84**, 869–70.
89. Bass-Shadkhan, Kh. F. (1962). Mikroelementy v Sel'. Khoz. i Med. Ukr. Nauch.-Issled. Inst. Fiziol. Rast., Akad. Nauk Ukr. S.S.R., Materialy 4-go (Chetvertogo) Vses. Soveshch., Kiev, 640–3; (1965). *C.A.* **63**, 13733.
90. Vulfs, L. and Vilks, S. (1968). *Latv. PSR Zināt. Akad. Vest.* No. 8, 98–104; (1968). *C.A.* **69**, 104043.
91. Khrycheva, A. I. (1970). *Prikl. Biokhim. Mikrobiol.* **6**, 307–12; (1970). *C.A.* **73**, 85023.
92. Koval'skii, V. V. and Letunova, S. V. (1964). *Usp. Sovrem. Biol.* **57**, 71–89; (1964). *C.A.* **60**, 16229.
93. Reisenauer, H. M. (1972). *Atti Simp. Int. Agrochim.* **9**, 533–8.
94. Tyagny-Ryadno, M. G. and Zvachkina, A. A. (1961). Rol Mikroelementov v Sel'. Khoz., Trud. 2-go (Vtorogo) Mezhvug. Soveshch. po Mikroelementam, 246–51; (1962). *C.A.* **57**, 11581.
95. Welkie, G. W. (1968). *Biochim. Biophys. Acta* **158**, 479–83.

6 Effect of Cobalt on Enzymes

In view of the catalytic behavior of cobalt in the following processes: the production of liquid hydrocarbons from carbon monoxide and hydrogen; the hydrodesulphurization of petroleum stocks; the addition of carbon monoxide and hydrogen to an olefin to form aldehydes and primary alcohols; in driers for paints, varnishes, printing inks, etc.; and in many other important industrial reactions, it was inevitable that numerous investigators should examine the effect of cobalt on the most important biological catalysts—enzymes.

Because the substances known as enzymes play a very important role throughout the whole of organized nature, brief references to a few of them have already been made in previous chapters. At the risk of a minor amount of repetition, however, it was believed desirable to summarize in this chapter information on the role of cobalt in enzyme activity. The enzymes will be discussed in alphabetical sequence.

Mold acylase is activated by cobalt, and the results suggest that mold acylase has a metal which is firmly bound to protein and necessary for enzyme activity.[1] The addition of cobalt may contribute to the removal of any amino acids formed. Cobaltous ion augmented the hydrolytic rate of acetyl-L-leucine by mold acylase, whereas it decreased that of chloracetyl-L-phenylalanine.[2]

Crystalline bacterial α-amylase was inactivated by incubation with a number of cobalt complexes.[3] Cobalt chloride added to human saliva increased amylolytic activity, but cobalt nitrate had no significant effect on salivary amylase.[4] Cobalt chloride, and, to a lesser extent, the nitrate and sulphate, increased the activity of pancreatic amylase.

A cobalt concentration of 2×10^{-3} M reduces the catalase activity in liquid cultures of the bacterium *Proteus vulgaris*[5] and of the fungus *Neurospora crassa*.[6] In the latter, the strong inhibition caused by cobalt when nitrate was included in the medium appeared to be attributable to more than simply the iron deficiency produced.

Dehydrogenase activity of lupine *Azotobacter* and *Rhizobia* cultures was increased by 0·001–0·01 mg cobalt l^{-1};[7] the enzymes in root nodules of beans and soybeans were also stimulated by cobalt.[8] Dehydrogenase activity of *Rhizobia* was increased 40% by 0·01 mg cobalt l^{-1}, but 0·1–50 mg l^{-1} caused an inhibition.[9]

The addition of a small quantity of cobalt caused a nine-fold increase in the activity of the alkaline β-glycerophosphatase in the small intestine of the rabbit.[10] Cobalt concentrations of 0·25–40 parts 10^{-6} partially or completely inhibited the biosynthesis of glucanase.[11]

Hydrogenase in legume nodules was highly activated by cobalt;[12] it has been suggested that cobalt is bound to the enzyme and takes part in electron and hydrogen transfer of the oxidation–reduction reactions.[13]

Injections of 0·05–1 mg cobalt kg^{-1} did not affect serum lipase activity in rabbits, but 2–5 mg kg^{-1} increased the enzyme activity by about 10%.[14] Additions of 0·25–40 parts 10^{-6} cobalt partially or completely inhibited the formation of cell-lytic enzymes in cultures of *Actinomyces*.[11] The chloride, nitrate, and sulphate of cobalt increased the activity of human pancreatic lipase.[4]

Oxidase is inhibited by potassium cyanide, but the addition of cobalt chloride causes partial reactivation, possibly due to interaction of cobaltous ion with free cyanide.[15] Oxidase activity in *Rhizobia* was stimulated most at additions of 0·01–0·1 mg cobalt l^{-1}.[9]

A concentration of cobalt not exceeding 10 parts 10^{-6} stimulated phosphatases.[16] Another study showed that 0·3–0·5 mg cobalt l^{-1} enhanced slightly the activity of acid phosphatase in bakers' yeast, but that of alkaline phosphatase by 35–55%.[17] In a liquid-medium culture of *Actinomyces*, 0·25–40 parts 10^{-6} cobalt partially or completely inhibited the formation of protease,[11] but in a *Bacillus mesentericus* culture cobalt stimulated the synthesis of proteases.[18] At a concentration of 5×10^{-3} M, cobalt promoted the enzyme activity of *Pseudomonas aeruginosa* alkaline proteinase.[19]

Nitrate reductase formation in *Azotobacter vinelandii* and other organisms depended on cobalt or vitamin B$_{12}$,[20] and a deficiency of cobalt markedly reduced nitrate reductase activity in *Rhizobium*.[21] Small quantities of cobalt stimulated the nitrate reductase activity in leaves and nodules of legumes, but higher amounts caused chlorosis and a decrease in enzymic activity.[22]

Cobalt porphyrin synthase in extracts of *Clostridium tetanomorphum* inserts the cobaltous ion into many dicarboxylic porphyrins.[23]

The biosynthesis of zymosan in the walls of yeast cells was increased by the addition of cobalt to the nutrient medium,[24] and the presence of cobalt enhanced the immunobiological activity of zymosan produced by cells of *Saccharomyces cerevisiae*.[25]

Finally, it has been suggested that the effect of cobalt on the growth of *Brucella abortus* is due to an inhibition of synthesis of a respiratory enzyme,[26] and a study of enzyme properties of bacteria in metal-containing media

48   COBALT IN BIOLOGY AND BIOCHEMISTRY

showed a decrease in oxidative activity for pyruvate, lactate, succinate, and fumarate in a cobalt concentration of 5×10^{-4} M.[27]

References

1. Hata, T., Doi, E., and Asao, T. (1961). *Koso Kagaku Shinpojiumu* **15**, 6–11; (1961). *C.A.* **55**, 24856.
2. Doi, E. and Hata, T. (1963). *Mem. Res. Inst. Food Sci., Kyoto Univ.* No. 25, 25–33; (1964). *C.A.* **60**, 3238.
3. Pomeranz, Y. (1963). *Biochim. Biophys. Acta* **77**, 451–4.
4. Shkol'nik, M. I. (1965). *Uchen. Zap. Petrozavodsk. Gos. Univ.* **12**, 136–40; (1966). *C.A.* **65**, 2559.
5. Petras, E. (1957). *Archs. Mikrobiol.* **28**, 138–44; (1959). *C.A.* **53**, 16273.
6. Subramanian, K. N. and Sarma, P. S. (1968). *Biochim. Biophys. Acta.* **156**, 199–202.
7. Koleshko, O. I. (1970). *Fiziol Biokhim. Mikroorg.* 177–81; (1972). *C.A.* **76**, 1642.
8. Yagodin, B. A. and Ovcharenko, G. A. (1969). *Izv. Akad. Nauk S.S.S.R., Ser. Biol.* No. 1, 113–21; (1969). *C.A.* **70**, 86659.
9. Koleshko, O. I. and Danil'chik, N. I. (1966). Mikroelem. Sel'. Khoz. Med. Dokl. Vses. Soveshch. Mikroelem. 5th, 474–8; (1970). *C.A.* **72**, 129613.
10. Clark, B. and Porteous, J. W. (1963). *Biochem. J.* **89**, 100.
11. Shukan, L. A. and Shklyar, B. Kh. (1975). *Biol. Akt. Veshchestva Mikroorg.*, 44–9; (1976). *C.A.* **85**, 119287.
12. Peive, Ya. V. (1967). *Izv. Akad. Nauk S.S.S.R., Ser. Biol.* No. 1, 11–19; (1967). *C.A.* **67**, 2422.
13. Peive, Ya. V., Yagodin, B. A., and Popazova, A. D. (1967). *Agrokhimiya* No. 1, 94–9; (1967). *C.A.* **66**, 104379.
14. Kichyna, M. M. and Shpakouski, A. U. (1967). *Vest. Akad. Navuk Belaruss. S.S.R., Ser. Sel'skagaspad. Navuk* No. 2, 79–82; (1967). *C.A.* **67**, 10640.
15. Tarkowski, S. (1966). *Medycyna Pr.* **17**, 116–19; (1967). *C.A.* **66**, 62161.
16. Niebroj, T. K. and Kozubska-Niebroj, M. (1964). *Acta Histochem.* **19**, 337–42; (1965). *C.A.* **62**, 13444.
17. Peciulis, J. (1967). Nauji Laimejimai Biol. Biochem. Liet. TSR Jaunuju Mokslininku-Biol. Biochem. Moksline Konf., 332–5; (1969). *C.A.* **71**, 110155.
18. Ciurlys, T., Uzkurenas, A., and Povilaitiene, J. (1975). *Liet. TSR Mokslų Akad. Darb., Ser. C* No. 3, 135–46; (1975). *C.A.* **83**, 160425.
19. Morihara, K. and Tsuzuki, H. (1974). *Agric. Biol. Chem.* **38**, 621–6; (1974). *C.A.* **81**, 22552.
20. Nicholas, D. J. D., Kobayashi, M., and Wilson, P. W. (1962). *Proc. Nat. Acad. Sci. U.S.* **48**, 1537–42.
21. Nicholas, D. J. D., Maruyama, Y., and Fisher, D. J. (1962). *Biochim. Biophys. Acta* **56**, 623–6.
22. Yagodin, B. A., Ovcharenko, G. A., Vasil'eva, Yu. Y., and Ivanova, M. A. (1970). *Sel'.-khoz. Biol.* **5**, 134–6; (1970). *C.A.* **73**, 34329.
23. Porra, R. J. and Ross, B. D. (1965). *Biochem. J.* **94**, 557–62.
24. Bass-Shadkhan, Kh. F. and Galina, I. (1961). *Latv. PSR Zināt. Akad. Vest.* No. 12, 69–74; (1962). *C.A.* **56**, 14729.

25. Bass-Shadkhan, Kh. F. (1962). Mikroelementy v Sel'. Khoz. i Med. Ukr. Nauch.-Issled. Inst. Fiziol. Rast., Akad. Nauk Ukr. S.S.R., Materialy 4-go (Chetvertogo) Vses. Soveshch., Kiev, 640–3; (1965). C.A. **63**, 13733.
26. Altenbern, R. A., Williams, D. R., and Ginoza, H. S. (1959). J. Bact. **77**, 509.
27. Kitamura, M. (1959). J. Okayama Med. Soc. **71**, 3681–88; (1960). C.A. **54**, 25054.

7 Cobalt in Plants

From the middle of the nineteenth century, cobalt was detected in plant material by the occasional investigator; the first systematic studies were published by Bertrand and Mokragnatz in 1930.[1] A few years later it became evident that cobalt deficiencies in livestock existed in many localities throughout the world, and investigations into the cobalt content of soils and plants increased rapidly. These examinations were greatly facilitated by the introduction at this period of photoelectric colorimeters, and the development of improved analytical methods for the determination of very small quantities of cobalt. Around 1960, the rapid extension of the technique of atomic absorption spectroscopy provided the analytical chemist with another valuable procedure for rapidly determining low concentrations of cobalt.

In Table 4 are given data published in recent years on the cobalt content of a number of plant materials, exclusive of hays and pastures, which are tabulated in Table 5, and of trees and shrubs, which are presented in Table 6.

Table 4 Cobalt Content of Plants

Plant	Cobalt (mg kg^{-1})	Ref.
Alfalfa	0·01–0·62	2
	0·02–0·84	3
	0·097–0·116	4
	0·2	5
	0·3	6
	0·24	7
	0·12–0·57	8
	0·28	9
	0·05–0·22	10
	0·04–0·29	11
	0·20–0·50	12
Alfalfa meal	0·282	13
Alfalfa silage	0·305	14
Apple, flowers	0·03	15
Apple, fruit	0·004	15
Apricot	0·03	1
Apricot, dried leaves	0·40	1
Astragalus	2·3–100	16

Table 4 (cont'd)

Plant	Cobalt (mg kg^{-1})	Ref.
Bahia grass	0·08	17
Baker's yeast	1–19	18
Bananas	0–0·026	13
Barley	0·1–0·18	19
	0·08	20
	0·13	9
Barley hay	0·06–0·11	21
Barley straw	0·24	7
Bavto straw	0·6–0·74	22
Beans	0–0·015	13
	0–1·8	23
	0·3	5
	0·70	24
Beans, edible	0·109	20
Beans, field, seed	0·01	1
Beans, French	0·07	25
Beans, horse	0·209	20
Beans, plant	0·1–1·6	26
Beans, wax	0·10	27
Beets, fodder	0·155	14
Beets, roots	0·05–0·09	27
Beets, tops	0·39–0·41	27
Bermuda grass	0·04–0·15	17,28
Bread, black	0·216	13
Bread, white	0·218	13
Brome grass	0·03–0·09	17,29
	0·2	6
Buckwheat, grain	0·36	1
	0·1	30
Cabbage, edible portion	0·07–0·24	1,25,27
Carolina blue grass	0·04	28
Carpet grass	0·05–0·15	17,28
Carrot, leaves	0·31	1
Carrot, roots	0·02	1
	0·006	13
Cassava	0·009	13
Cedar nuts	0·068–0·093	31
Centipede grass	0·04	28
Chard, Swiss	0·09	27
Cherries, edible portion	0·005	1
Clover, aloyce	0·09	28
Clover, red	1·3	32
	0·02–0·27	11
	0·1	33
	0·09–0·39	34
	0·37	9

Table 4 Cobalt Content of Plants (cont'd)

Plant	Cobalt (mg kg^{-1})	Ref.
Clover, subterranean	0·05–0·63	35
Clover, white	0·17–4·6	29,36
Coffee bean	0·002	1
Corn	0·52	33
Corn bread	0·159	13
Corn flour	0·185	37
Corn grain	0·01–0·02	1,25,27,28
	0·098	20
Corn silage	0·092	37
Corn silage, dry	0·192	14
Cowpeas	0·06–0·31	27
Cowpea hay	0·05–0·12	28
Cress, water	0·15	1
Dallis grass	0·03–0·15	17,28
Duckweed	0·48	38
Eggplant	0·031	13
Figs	0·20	1
Grapes, dry	90–110	39
Horseradish	0·121	40
Johnson grass	0·08	17
Jujube fruit	0·22–0·35	41
Kentucky bluegrass	0·13–0·25	17,29
Kodra straws	0·6–0·74	22
Lespedeza	0·03–0·73	28,42
Lettuce, dried leaves	0·05–0·23	1,27
Lime, dried leaves	0·20	1
Linseed	0·047	37
Lupine silage	3·44	37
Mangel-beet leaves	0·16–0·54	27
Molasses, sugar beet	1–9	18
Mushrooms	0·133–1·0	43
Natal grass	0·05	17
Oat bran	0·01	1
Oat chop	0·02	44
Oat flour	0·09	37
Oat forages	0·01–0·47	45
Oat hay	0·02–0·07	37
Oat straw	0·594	46
Oats, grain	tr.	1
	0·06	9
Oats, green	0·091–0·20	47
	0·39	22
Oats, whole plant	0·03–0·23	28
Onions, bulb	0·13	1
Onions, green	0·26	25
Orchard grass	0·08	17

Table 4 (cont'd)

Plant	Cobalt (mg kg^{-1})	Ref.
Paddy straw	0·031–0·094	48
	0·60–0·74	22
Para grass	0·07	17
Peanut hay	0·08	28
Pear, pulp	0·18	1
Peas, Australian field	0·15	28
Peas, dried	0·348	13
Peas, green, edible portion	0·03–0·15	29,49
Plankton, Black Sea	10	50
Plantain	0–0·026	13
Pepper, Guinea	0·02	13
Potatoes, ordinary	0·04–0·06	1
	0·003–0·079	13
	0·06	5
	0·35	9
	0·1	37
Potatoes, sweet	0·02–0·03	27
Quack grass	0·09	17
	1·17	6
Radish	0·30	25
Rape	0·10–0·12	49
Rape, winter	0·34	9
Red top	0·08	17
Rice, paddy	0·56–1·40	12
Rice, polished	0·006–0·13	1,51
Rice, straw	0·42–9·3	32
Rye	0·12	9
	0·087	20
Ryegrass, Italian	0·02–0·15	52
Salt	0·006	13
Sea grass	0·15–0·66	53
Seaweed	0·22–0·64	51
Sedges	0·6	9
Seradella	0·71	9
Slough hay	0·015	44
Sorghum	0·15–0·34	12
Sorghum, grain	0·008–0·01	45
Sorghum, silage	0·02–0·13	45
Soybean hay	0·05	28
Spinach, edible portion	0·07–1·20	1,27
Spinach, New Zealand	0·09	27
Strawberries, wild	0·10–0·12	49
Sugar beet	0·2	5
Sugar beet, cuttings	0·48	24
Sugar beet, leaves	0·43	9
Sunflower, green	0·13–0·27	47

Table 4   Cobalt Content of Plants (cont'd)

Plant	Cobalt (mg kg^{-1})	Ref.
Timothy hay	0·01–0·08	17
	0·053	54
Tobacco	0·28–0·62	55
Tobacco leaves	1·86	9
Tomato fruit	10	1
Turnip greens	0·03–1·07	27
Vasey grass	0·08	17
Vetch	0·3–0·35	28
Vetch, green	0·26	47
Walnuts, edible portion	0·05	1
Watermelon	0·18	49
Wheat	0·11	20
	0·05	5
	0·1–0·18	19
	0·058	56
Wheat, spring hard	0·140	57
Wheat, winter	0·082	57
	0·36	9
Wheat bran	0·01	1
Wheat flour	0·09	51
Wheat grain	0·01–0·04	1,25
	0·008–0·067	58
	0·34	7
Wheat straw	0·594	46
Wheat, whole plant	0·05–0·15	28
Wheat grasses	1–2	6
Wire grass	0·03	28
Witchgrass straw	0·06–0·11	21
Wormwood, pink-blossomed	0·874	59
Yams	0·012	13

It is evident from Table 4 that plant foods for animals and man vary markedly in their cobalt content. Differences occur not only between dissimilar plants, but also between the same species grown in different environments. The quantity of cobalt found in plants is dependent not only on the amount of this element in the soil, but also on a number of factors influencing availability and absorption of mineral nutrients. These include type of soil, moisture, aeration, temperature, soil pH, variety of plant, soil microflora, colloidal content, presence of other ions in soil solution, stage of plant growth, soil drainage, organic matter, form of cobalt and other variables.

The principal categories of Table 5, fodders and forages, grass, hay, and pasture herbage, do not represent individual species and would be expected

Table 5 Cobalt Content of Hays and Pastures

Hay and pasture	Cobalt (mg kg^{-1})	Ref.
Cereals	0·6	9
Cereal straws	0·64	12
Feeds and forages, cobalt-deficient	0·04–0·07	42
Fodders	0·20	60
	0·25–1·4	61
Fodders, low in cobalt	0·030–0·162	62
Fodders, green, dried	0·24–0·25	63
Forages	0·1–0·5	64
	0·067	65
	0·03–0·10	66
Forages, untreated	0·026–0·048	67
Forages, treated with cobalt	0·082–0·121	67
Forage, green grass	0·042–0·31	47
Grass	0·22	9
	0·2–1	68
, max. of 284 Japanese samples	0·38	69
	0·36	12
	0·36	70
	$\lesseqgtr 0.39$	71
Grass, pasture	0·02–0·06	72
	0·24–0·25	63
	0·1–0·2	73
	0·073–0·203	74
	0·25	75
Grass, low cobalt	0·058	76
Grass, spring	0·35	77
Grass, wild, Japan	0·164	78
Hay	0·03–0·19	11
	0·15	79
	0·15	63
	0·16–0·2	80
	0·02–0·04	81
	0·04–0·38	82
	0·233	46
	0·02	5
	0·11–0·79	83
	0·14–0·21	84
	0·24	85
	0·071–0·093	4
Hay, meadow	0·07–0·42	86
	0·02–0·25	87
, av. of 200 analyses	0·14	88
Hay, marshland, first cutting	0·098	37
Hay, marshland, second cutting	0·103	37
Hay, mineral soil	0·073	37

Table 5   Cobalt Content of Hays and Pastures (cont'd)

Hay and pasture	Cobalt (mg kg^{-1})	Ref.
Herbage, pasture,		
Australia	0·02–0·43	2
Britain	0·01–0·40	2
Canada	0·02–0·03	2
Germany	0·02–0·38	2
India	0·20–1·60	2
Italy	0·02–0·62	2
New Zealand	0·01–1·26	2
Norway	0·03–0·25	2
Poland	0·04–0·26	2
Switzerland	0·07–0·42	2
Uruguay	0·05–0·76	2
U.S.A.	0·01–0·73	2
U.S.S.R.	0·12–0·58	2
Herbage, pasture	0·4–1·1	89
	0·05–0·60	90
Herbage, pasture, av.	1·6	91
Herbage, pasture, desert and semi-desert	0·39	77
Herbage, pasture, sandy desert	0·1–0·7	92
Legumes	0·8	9
	0·4–1·6	68
	0·035–0·087	93
Legumes, low cobalt	0·11–0·15	76
Plants,		
Caryophyelaceae	0·19	94
Chenopodiaceae	0·30	94
Compositeae	0·42	94
Gramineae	0·50–3·01	95
Gramineae	0·40	94
Leguminosae	0·28–1·31	94
medicinal	0·02–0·31	96
medicinal	0·07–0·58	97
normal	0·6–1·01	98
normal	0·35–0·80	99
Plants, normal, soddy alluvial soil	0·47	100
Produce, green	12	101
Silage, dry residue	0·091–0·098	4
Silage, lupine	3·44	37
Silage, maize	0·09	37
Silage, sorghum	0·02–0·13	45
Weeds, meadow	0·8	9
Wheat grass, *Agropyron desertorum* and *A. intermedium*	20	102

Table 6 Cobalt Content of Trees and Shrubs

Trees and shrubs	Cobalt (mg kg^{-1})	Ref.
Alder	0·06–0·25	103
Arbutus	5·67	104
Balsam	0·1–0·6	103
Beech, bark	1·10	1
Beech, dried leaves	0·35	1
Beech, leaves	0·03–0·70	105
Beech, wood	0·01	1
Beech, wood	5·2	106
Birch	0·03–0·25	103
Black gum, leaves	5	107
Cedar	0·1	103
Douglas fir	0·40	104
Douglas fir, needles	0·05–0·46	108
Hornbeam	5·2	106
Horse chestnut leaves	0·03–0·70	105
Jackpine	0·08–0·1	103
Labrador tea	0·04–0·75	103
Lodgepole pine	0·25	103
Lodgepole pine, needles	0·05–0·46	108
Maple	0·08	103
Mountain balsam	0·3	103
Oak	5·2	106
Pin cherry	0·02	103
Poplar	0·1–0·3	103
Sycamore leaves	0·03–0·70	105
Tamarack	0·06–0·15	103
Western hemlock, needles	0·05–0·46	108
Willow	0·08–0·35	103
Willow bark, low in cobalt	0·066	109

to be more uniform than the plants of Table 4. This is indeed so, in spite of a few high values recorded. Fodders and forages with a low cobalt content of 0·02–0·05 mg kg^{-1} are cobalt-deficient, whereas most contain 0·1–0·25 mg kg^{-1}. Low-cobalt grass has 0·02–0·06 mg kg^{-1} against the usual average of 0·2–0·35 mg Co kg^{-1}. Some hays are low in cobalt with 0·02–0·04 mg kg^{-1}, but most contain about 0·1–0·2 mg kg^{-1}. Deficient pastures may have 0·01–0·05 mg Co kg^{-1}, but the majority of pasture herbage contains 0·1–0·35 mg kg^{-1}.

The values for cobalt in trees and shrubs, in Table 6, illustrate the wide variation in the uptake of this element by different members of the plant kingdom under dissimilar environments. It should be mentioned that the data in Table 6 are for trees and shrubs in areas where no cobalt mineralization was apparent. They would represent the normal background values in geochemical prospecting, whereas samples from mineralized areas might contain 10–100 times these quantities.

Some trees appear to concentrate cobalt more than others. Spruce and birch tend to accumulate more cobalt than willow and maple when all are in close association. The arbutus and Douglas fir cited in Table 6 were growing side by side, yet the arbutus accumulated 14 times the amount of cobalt found in the fir.[104] Several workers have reported that the black gum, *Nyssa sylvatica*, accumulated a far greater amount of cobalt in mature foliage than did other species on the same soil.[110–112] Another tree, *Clethra barbinervis*, contained more than 100 times as much cobalt in the leaf ash as did other plant species growing in the same vicinity.[113]

Cobalt-deficient Fodders and Forages

We have seen in the previous chapter that in many parts of the world the soils, and the plants growing on them, do not provide sufficient cobalt for the maintenance of health in cattle and sheep. In forages and pastures, the minimum quantity of cobalt required to prevent deficiency diseases has been given by a number of workers as 0.07 mg kg^{-1} of dry feed.[114–120] In Scandinavia, a slightly higher value has been proposed, for cattle, of 0.1 mg kg^{-1}.[121] In South Africa, sheep on pastures containing $0.06 \pm 0.04 \text{ m Co kg}^{-1}$ remained healthy.[122] A wider range is given by a Brazilian worker: unhealthy forage at $0.046 \text{ mg Co kg}^{-1}$ and healthy forage at 0.334 mg kg^{-1}.[123] Higher levels have been recommended by some Russian investigators: 0.3 mg kg^{-1} dry feed for sheep[124] and $0.30–0.37 \text{ mg kg}^{-1}$ for a ewe with one lamb,[125] $0.7–0.8 \text{ mg Co kg}^{-1}$ feed for calves,[126] $1.2–1.7 \text{ mg kg}^{-1}$ for cows,[127] and 2 mg Co kg^{-1} of dry feed for calves of age 4–6 months.[128]

Many publications confirm the common occurrence of low cobalt levels in feeds and pastures, in many parts of the world, which causes a cobalt deficiency in ruminants.[2,129–141] Cobalt has been found to be adequate in some forage plants of western India.[142] There is a review of cobalt and other trace elements in sheep pastures and fodder crops in Australia.[143]

Uptake of Cobalt in Plants

There have been numerous studies on the uptake of cobalt in plants. These have included the relation between natural soil content of cobalt, or cobalt additions to soil, and the concentration of this element in the plant, the variation of cobalt content with plant species, changes in cobalt with growth, differences in concentration of cobalt in various parts of a plant, and the influence of lime, soil acidity, organic matter, and fertilizers on cobalt uptake in plants.

As a general rule, plants growing on a soil or other medium which has a low cobalt content will be low in cobalt. The application of cobalt salts or other cobalt-containing materials increases the cobalt content of plants.[30,144-155]

Many observations have been reported on the variation of cobalt uptake by different plant species. Most workers found cobalt higher in legumes than in grasses or cereals,[86,156-159] but for wild plants of pastures, one worker found a range of 0·28–1·31 mg Co kg^{-1} of dry plant for legumes, and 0·50–3·01 mg kg^{-1} for grasses.[95] It has been reported that grasses and sedges contained less cobalt than weeds,[160] and that cereals consumed about 2 % of the available cobalt in the soil.[161] It was noted that winter rye, timothy, and potatoes took up the same amount of cobalt from the soil.[162]

The distribution of cobalt in the different parts of plants has been studied by numerous workers; the radioactive tracer, ^{60}Co, has been extensively used in such investigations. In most plants, cobalt has been found to accumulate mainly in the roots;[163-167] one paper reported that cobalt accumulated in alfalfa roots, but in corn and cotton seeds.[168] It was observed that the cobalt content of seeds of a number of vegetables and fruits was 2–3 times higher than in their edible parts.[40] Other workers have noted that in barley the grain is relatively low in cobalt,[169] the level of cobalt in green feeds is higher than in grain from the same plant,[170] and in peas and oats a higher concentration of cobalt is found in the whole plant than in the foliage alone.[171] Investigators have reported that tree barks are 5–7 times richer in cobalt than are woody tissues,[172] that birch leaves contained 2–3 times more cobalt than the trunk,[173] and that young sprouts on the top of trees contained less cobalt than those in the middle parts of the trees.[174]

A few studies have probed into the changes in cobalt content of plants throughout their growth period. As a general rule, the cobalt concentration increases during the growth period, then decreases after flowering.[175-178] Usually the second crop of pasture grass has a lower cobalt content than the first cutting.[179]

The effect of organic matter on the uptake of cobalt in plants has been investigated. It has been found that more cobalt was picked up by plants in light soils with a low organic matter content than was the case in soils rich in organic constituents.[180] Perhaps this has been corroborated by a paper which found that the application of 20 tons of manure per hectare gave, for a variety of crops, a lower cobalt content than the control.[181] In a study on the solution of metal oxides by decomposing plant materials, it was reported that dissolved cobalt is in a complex form in true solution and does not undergo ion exchange to any extent.[182] In sand cultures, the highest uptake of cobalt was observed from organic complexes obtained from aerobic fermentation, whereas in water cultures a cobalt–anaerobic solution or a cobalt–mineral solution were more effective.[183]

Several workers have reported lower cobalt levels in plants grown on either calcareous rocks or limed soils.[184–188] A related conclusion indicated that plants on alkali soils showed a lower cobalt content than the same plants grown on other soils.[187] Another corroboration comes from a paper which showed higher cobalt in plants from acid soils, in the pH range 4·9–6·2.[189] Large increases in the uptake of cobalt by Sudan grass were found when ferrous chloride, aluminium chloride, and calcium chloride were added to the soil.[189] In a study of a soil–vegetative complex in Russia, cobalt was found only in plants grown on peat-boggy soils.[190]

On the other hand, one worker has reported that cobalt uptake was enhanced by lime and ammonium sulphate fertilization.[191] It is probable, however, that the depressing action of lime was overcome by the presence of ammonium sulphate. A study on the effect of fertilizers on absorption of trace elements by plants comes to the very sensible conclusion that crops grown on fertilized soil were capable of using to greater advantage the merits of cobalt and other trace elements present in soils.[192] Australian workers have indicated that plants in soil containing more than 1000 mg of manganese per kg of soil are unlikely to respond to cobalt fertilizers added to the soil.[193] It has been reported that, when a large quantity of lead is present in a soil, cobalt present in pastures reaches its lowest value in summer, when lead content is maximum.[194]

Effect of Cobalt on Plant Yield

If it has not been demonstrated conclusively that cobalt is an essential element in plant nutrition, at least there is abundant evidence that the addition of a small quantity of cobalt to the soil, or other growth medium, usually results in many plants having an increased yield. Over many years, a large number of investigations in various countries have reported a beneficial effect on plant yield from small additions of cobalt; a few workers, however, have failed to find such improvement.

In experiments designed to determine the effect of cobalt on crop yield, addition of cobalt is made in four ways:

1) addition of solid, water-soluble cobalt compounds to the soil, sand, or water culture;
2) soaking the seed or tuber, prior to sowing, in a dilute solution of a water-soluble cobalt compound;

3) spraying the plant foliage at an early stage with a water-soluble cobalt compound;
4) dusting the seeds, before sowing, with a finely powdered water-soluble cobalt compound. (This practice is not nearly as common as the first three methods listed.)

The most popular procedure is the addition of cobalt sulphate, nitrate, or chloride to the soil, sand culture, or water culture. The concentrations employed cover a wide range, $0.001–120$ mg kg^{-1}, but most have been about $0.5–5$ mg Co kg^{-1} of soil, or its equivalent. Many experiments with this method report the additions as kg Co ha^{-1}. The latter is probably very useful from an agronomic standpoint, but makes it impossible to compare treatments on a strictly mathematical basis. The assumption has often been made that a hectare of soil to a depth of 15 cm weighs about 2 240 000 kg; this would mean 1 kg ha^{-1} = about 0.45 mg kg^{-1}. In view of the considerable variation in soils, however, such figures must be viewed as very rough approximations.

In recent years, the addition of cobalt by presoaking seeds or tubers in cobalt sulphate, nitrate, or chloride solution for 2–24 h, usually 4–12 h, has been used by many workers. The range of cobalt concentrations has been wide, $0.0002–0.6\%$ as the cobalt compound, but usually about 0.05% of cobalt sulphate, nitrate, or chloride is employed. The sulphate, nitrate, and chloride of cobalt are hydrated salts containing, respectively, $20.96, 20.25$, and 24.77% of cobalt.

A third method of adding cobalt to crops is by foliar fertilization, that is, spraying the young plants with a dilute solution of a cobalt salt. The sulphate or nitrate of cobalt are almost invariably used; the concentration may range from about $0.005–0.2\%$, generally being around 0.05%. Unless great care is exercised, foliage spraying will result in a cobalt addition to the soil, and it becomes difficult to assess whether the crop yield has been affected by foliar or soil fertilization.

Finally, a few experiments have been reported on the effect of dusting seeds, prior to sowing, with finely powdered cobalt sulphate or nitrate. The addition of the cobalt salt has been made at rates of 20–400 mg kg^{-1} of seed, with a rough average of 100 mg kg^{-1}. The advantage of either soaking or dusting the seed over applying cobalt to soil is that, in the former, traces of cobalt are presumably in direct contact with the germinating seed or sprouting tuber, and are thus immediately available to the plant. On the other hand, addition of cobalt compounds to the soil instead of the seeds should result in a better distribution of the element in the growth medium, thereby providing a larger reserve for the entire growth period of the plant.

In one of the earliest investigations on the effect of cobalt on plant yield, a beneficial effect on algae was obtained at 0.2 parts 10^{-6} in water culture; for timothy in soil, the lowest cobalt concentration of 0.1 mg kg^{-1}, or parts 10^{-6}, gave a yield approximately equal to that using untreated soil, higher concentrations being toxic.[195] Brief reference to the relation between cobalt and plant yield has been included in many reviews on cobalt and other trace elements in both plant and animal nutrition.[2,196-210] Recent work in this field will be examined in succeeding paragraphs. In many cases, an increase in yield was accompanied by an improvement in the quality of the crop, and, if so, this result will be briefly recorded later in this section. A detailed review of the effect of cobalt on specific aspects of quality, i.e. protein or carbohydrates in cereals and foodstuffs, sugars in sugar beets, etc. will be covered in a subsequent section.

Among cereals, the yields of barley, buckwheat, and oats were increased by soil additions of cobalt;[211-215] other workers did not obtain an increase in yield of either wheat[216,217] or oats.[218] Cobalt fertilization by presoaking seeds has been reported to increase yields of barley, buckwheat, wheat, and oats.[219-223] Foliar fertilization improved the yield of both wheat and millet.[224]

Yields of corn have been invariably increased by additions of cobalt to the soil;[225-231] soaking seeds in cobalt solutions[232-236] and the use of foliar fertilization[237] also proved beneficial to this crop.

Both cobalt additions to the soil[238-244] and presoaking seeds in cobalt solution[245,246] improved the yield and quality of cotton.

The yield and quality of flax have been improved both by cobalt additions to the soil,[247,248] and by the use of solutions to soak the seeds prior to sowing.[249]

Cobalt additions to soil,[250-252] and foliar spraying with dilute cobalt solutions,[253] have increased the yield and improved the quality of grapes.

Timothy in Russia and Kentucky 31 fescue in Japan were improved by cobalt additions to the soil,[254,255] but grass herbage in the province of Nova Scotia in Canada showed no increase in yield with similar treatment.[256]

Many publications have appeared on the influence of cobalt on the yield and quality of legumes; the majority have reported beneficial effects from additions of this element to soils.[257-272] One paper reported a decreased yield, and stated that the addition of cobalt to a heavy clay soil is not to be recommended;[273] another study found cobalt was beneficial only in the presence of conventional fertilizers.[274] All workers agree on a beneficial effect from soaking legume seeds in a cobalt solution;[275-285] the one paper on the use of foliar spraying of legumes indicates an increase in yield.[286] In concluding the subject of cobalt and legumes, studies have indicated that cobalt is an essential element for growth of soybean plants when they depend on the symbiotic relation with *Rhizobium japonicum* for the fixation of atmospheric nitrogen;[287−289] other investigators have reported that the legume itself, as distinct from its symbiont, requires cobalt.[290]

Most investigators have found that cobalt additions to soil improved both the yield and quality of sugar beets;[291−295] one worker reported that cobalt usually has no effect.[296] Seed treatment;[292,294,296,297] foliar application of cobalt solutions,[292−294,298] and dusting[293,299] all had beneficial effects on the yield and quality of sugar beets.

Many experiments have been carried out on the effect of cobalt on the yield of vegetables. Additions of cobalt to the soil have been consistently beneficial.[300−311] Nearly all papers on the effect of presoaking either seeds or tubers show an increased yield;[312−322] one worker reported a decrease in yield.[323] In most trials, spraying foliage with a dilute cobalt solution gave beneficial results[323−328] but one paper reports a lack of improvement.[329]

A number of papers have reported favorable results on the growth of miscellaneous plants by additions of cobalt. Additions to soil have benefited *Lilium* pollen,[330] *Alnus rubra*,[331] chamomile flowers and peppermint leaves,[332] and *Datura stramonium*.[333] Soaking seed with a cobalt solution improved the growth of acorns,[334] and groundnuts;[335] it also increased the resistance of radish and rye grass to industrial pollutants such as sulphur gases.[336] Foliar spraying with cobalt solutions improved the growth of *Calendula officinalis*,[337] *Quercus robur* seedlings,[338] and European ash buds.[339]

Comparing the addition of cobalt to soil, spraying leaves with a cobalt solution, and wetting seeds with a cobalt solution, indicated that the latter treatment was more beneficial for sugar beets.[340] In an experiment with alfalfa, wetting seeds was found to be better than dusting.[341]

Effect of Cobalt on Chlorophyll and Photosynthesis

The addition of cobalt to the soil, or the presoaking of seeds in a dilute cobalt solution, have been found to increase the amount of chlorophyll in barley,[342−344] buckwheat,[213,220] and oats.[345] A similar benefit was noted in corn,[227] grapes,[252,253,346,347] legumes,[260,348,349] sugar beets,[350−356] vegetables,[301,316,320,327,357−364] trees,[331,365−367] and miscellaneous plants.[368−370]

Influence of Cobalt on Protein and Nitrogen Metabolism in Plants

Many investigations have revealed that the addition of cobalt to the growth medium has resulted in an increase in either nitrogen or protein content of plants. Among cereals, cobalt additions to soil, or seed treatment with a cobalt solution increased the protein in buckwheat,[215,221] oats,[218,371] and wheat.[217] The protein content of corn was increased by similar cobalt additions.[226,227,229,230,233,372] It was reported that wetting the seeds of cotton with 0·01% cobalt nitrate solution accelerated the hydrolysis of protein; a 0·5% solution resulted in inhibition.[373] Spraying grapevines with a dilute cobalt solution increased the protein content of the leaves between 3- and 5-fold.[374]

Numerous papers have shown that the addition of cobalt to legumes gave an increased protein content in clover,[217,263,272,375] lupine,[261,269,283] peas,[269,276,282,284,376,377] and legume hay.[378,379] On the other hand, an Australian worker found that cobalt had no effect on the nitrogen content of subterranean clover,[257] and Russian investigators failed to find any influence of cobalt on the protein content of either peas[267] or beans.[280]

The protein content of potatoes has been increased by soaking the tubers, before planting, in 0·05% cobalt nitrate solution.[308,322] Treating the seed and foliage of peanut with cobalt increased nitrogen content.[335] Cobalt additions to soil increased the total nitrogen and protein of the white mulberry tree.[380] The nitrogen content of sunflower leaves was augmented by spraying the foliage with a cobalt solution.[381] The addition of about 1 mg cobalt kg^{-1} of soil increased the proline and lysine content of plants, but not that of histidine.[382]

Effect of Cobalt on Sugars and other Carbohydrates in Plants

The influence of cobalt additions to the growth medium on the quantity of sugars or other carbohydrates in plants has attracted the interest of numerous investigators. Cobalt increased the starch content of wheat,[217] and of corn,[234] and also the sugar content of corn;[229] cobalt augmented sugar formation in timothy.[383]

The accumulation of sugar in grapes is favored by the addition of cobalt.[252,253,384,385] One worker reported that cobalt increased the sugar content of clover,[217] but another paper showed that there was no effect on carbohydrate content.[265] In lupine, the contents of starch[269] and of sugars[283] have been increased by cobalt applications to soil, seed, and foliage. Carbohydrates in peas,[284] and in peas and beans,[386] showed an increase with cobalt applications.

All investigators have reported an increased sugar content in sugar beets by additions of cobalt to soil, to seed-soaking solutions, foliar sprays, or to seed-dusting powders.[239,254,296-300] Cobalt improved the starch content of potatoes,[308,321,322,363] and in some examples, also the sugar of this crop.[328] The sugar content of tomatoes was increased by cobalt,[237,387] as was the carbohydrate content of the white mulberry tree.[380]

Effect of Cobalt on Enzymes in Plants

A number of publications have appeared reporting the effects of cobalt additions on enzyme activity in plants. In barley, cobalt depressed the development of glutamate-oxalacetate transaminase, but had no influence on the activity of either urease or amylase.[388] Cobalt was found to decrease the activity of dehydrogenases in barley,[389] but a later paper by the same worker reported an increase in dehydrogenase activity.[390] Catalase activity was raised by adding cobalt to cotton plants.[240]

Cobalt additions to grapevines increased the activity of catalase, ascorbic oxidase, polyphenoloxidase, and peroxidase.[347] In sunflowers, cobalt depressed catalase in young leaves but not generally in older leaves.[391] Cobalt increased the activities of both catalase and peroxidase in the meadow foxtail.[392]

In legumes, cobalt was found to exert a positive effect on the enzyme activity of nodules.[393,394] Cobalt additions to fodder beets increased the carboanhydrase activity but had no effect on aldolase;[395] in the same crop, ATPase activity was increased by additions of cobalt.[396]

In potatoes, the addition of cobalt enhanced catalase activity[397,398] and dehydrogenase activity.[399] Soaking tomato seeds in cobalt solutions prior to sowing increased the activities of both catalase and peroxidase.[400] One investigator has stated that cobalt is directly bound to enzymes that catalyse the fixation of nitrogen, the reduction of nitrates, and the synthesis of amino acids, nucleic acid, and proteins.[401]

Cobalt and Photoperiodism

The suppression of the elongation of bean hypocotyls by light was intensified moderately by use of a dilute cobalt solution.[402] Cobalt accelerated the photoperiodic phase in oats under short-day conditions.[403] In a study of lupines, under long-day conditions the number of nodules per plant was increased 1·7 times by cobalt; in short-day conditions it was augmented 2·7 times.[404]

Cobalt and Transpiration

Some investigations have been carried out on the effect of cobalt on transpiration in plants. Transpiration rates of both kidney beans and mustard plants improved with cobalt additions.[260] The transpiration rate of cabbage plants treated with cobalt sulphate was 1·5–2 times that of controls; cobalt chloride increased transpiration rates to a lesser extent.[405] In grapevines, cobalt intensifies transpiration when sufficient water is present and reduces it when there is insufficient water.[406] For potatoes receiving cobalt, maximum transpiration was observed during the morning hours, in the leaves of plants grown from tubers moistened with 0·05% solution of cobalt nitrate; the maximum for tubers treated with 0·25% solution occurred at noon.[407] In buckwheat, cobalt decreased the intensity of transpiration in hot, daytime periods.

The Relation between Cobalt and Lime in Plant Growth

It has been reported that applications of lime changed soluble cobalt compounds into less available forms, and additions of cobalt significantly increased both the yield and quality of crops in these limed areas.[286,292,408-411] Liming reduced the percentage of cobalt in hay from natural meadows.[378]

Cobalt and Translocation

In a study of carbohydrate translocation in the cotton plant, the addition of cobalt did not appear to have any effect.[412-414] An investigation into the translocation of cobalt in the tomato plant indicated that cobalt was transported predominantly as an inorganic cation in the stem exudate.[415] Cobalt, added either to soil or to solutions used for soaking seeds prior to sowing, increased the assimilate flow into the roots and root nodules of peas and beans during the flowering stage, and enhanced the assimilate translocation into the generative organs during the stage of seed formation.[416]

Cobalt and Drought-resistance of Plants

A few papers have appeared on the effects of cobalt on the drought-resistance of various plants. Cobalt increased the drought-resistance of corn and soybeans,[225] oats,[345] oats and barley,[417] and barley.[418] It has been reported that the effect of cobalt is related to its capacity to increase the content of bound water, maintain a high protein content, and to increase the rates of synthesis and migration of carbohydrates from the leaves to the fruit-bearing organs.[419] Another publication stated that, among trace elements, cobalt had the most effective action on drought-resistance, and increased the content of adenosine triphosphate.[420] The increase of drought resistance induced by cobalt has been ascribed to its direct or indirect participation in nucleic acid metabolism.[421]

Cobalt-accumulating Plants

Some plants appear to accumulate cobalt; the content of this element is much higher than in other species growing in the same location. *Astragalus* sp contain the very high concentration of 2·3–100 mg cobalt kg^{-1} dried plant material.[16] *Artemesia* or wormwood species are rich in cobalt;[59,422] pink-blossomed wormwood was found to contain 0·874 mg cobalt kg^{-1} of dry plant material, and sheep grazing on pastures where this grass predominates do not require supplemental cobalt.[59]

The wheat grasses *Agropyron desertorum* and *Agropyron intermedium* in Nevada have been reported to contain 20 mg cobalt kg^{-1} dry matter.[102] Sagebrush-legume herbage in Azerbaidzhan contains 4–5 times more cobalt than steppe grasses.[423] A study of a New Zealand serpentine flora indicated that *Pimelea suteri* was a strong cobalt accumulator.[424] A publication on hill herbage suggested that *Calluna* may accumulate cobalt, and *Trichophorum* and *Molinia* may reject it even if available in the soil.[425] Cobalt, like other microelements, may accumulate in some plants growing in regions of ore deposits;[426,427] samples from mineralized areas can have 10–100 times the quantity of cobalt found in unmineralized regions.[2]

In discussing cobalt in trees and shrubs earlier in this chapter, reference was made to the accumulation of cobalt in arbutus,[104] in black gum,[110–112] and in *Clethra barbinervis*.[113]

Toxicity of Cobalt in Plants

A few investigators have studied the toxic effects of cobalt on plants. In Italy, cobalt was toxic to oats at concentrations greater than 10–40 mg kg^{-1} of soil, in acid, basic, and neutral soils.[428]

In Australia, cobalt was present in the soil solution at a concentration of 0·03–0·14 parts 10^{-6}, or at less than 10 % of the level needed to induce toxicity in an oat crop.[429] The seedling tops of corn were injured by a cobalt content in the soil of 25 mg kg^{-1} of soil.[430] In rice plants, toxic levels of cobalt decreased glumes or ears.[431]

When seeds of broad beans were soaked in 0·1 % cobalt sulphate solution prior to sowing, 20% of the plants were chlorotic as a result of excessive cobalt; in the chlorotic leaves, decreased chlorophyll was accompanied by reductions of protein.[432] In another study of beans, it was reported that magnesium sulphate partially counteracted cobalt toxicity, whereas ammonium ferric citrate had no beneficial effect against excessive cobalt.[433]

Cobalt treatments of soybeans, either seed or foliar, greater than 36·8 mg of cobalt sulphate kg^{-1} of seed, produced chlorosis of the leaves and stunted the young seedlings; mild chlorosis could be completely corrected by adding iron salts.[434] Cobalt added to cabbage in water cultures at 0–5 parts 10^{-6} decreased the yield at high levels; the cobalt content in the outer leaves was about 30 mg kg^{-1} dry matter when yield decreased to 50%.[435] When chrysanthemums in water culture received cobalt additions of 10^{-6}–10^{-4} M, growth was depressed 45 % at 10^{-4} M; most of the cobalt was associated with the roots, followed by the leaves and the stems.[436]

Miscellaneous Effects of Cobalt on Plant Growth

A number of publications have appeared on the effect of cobalt on various unrelated aspects of plant growth. In the mulberry, addition of cobalt sulphate improved the absorption of phosphorus.[437] Another study on the same tree found cobalt additions increased the leaf area of seedlings, stimulated enlargement of root necks, and improved the frost resistance of seedlings.[438]

Investigations on cotton plants showed that a foliar spray of cobalt nitrate acted as a defoliant if applied when the bolls were opening,[439] and that the application of 100 kg $CoCl_2.6H_2O$ ha^{-1} was most effective in controlling verticillium wilt disease.[244]

Experiments on sunflowers revealed that cobalt additions to soil increased the fat content of seeds,[440] and in water cultures, the addition of cobalt produced roots in plants which had previously had their roots cut off.[441]

Either introducing cobalt into the soil or spraying the leaves with a cobalt solution enhanced the ascorbic acid content of tomato.[442] The effect of cobalt on the production of riboflavine by *Candida guilliermondii* has been shown to be due to a competition with iron, and produces the same effects as iron deficiency.[443]

Spraying *Calendula officinalis* with a 50 mg l^{-1} solution of cobalt nitrate greatly increased the content of carotene.[337] In corn, cobalt increased the transformation of anthranilic acid into glycosides.[444] Calcium absorption in beans was increased by cobalt additions.[262]

The application of 0·1–0·2% cobalt nitrate solution caused differentiation of the endoplasmic reticulum membrane complexes in horse beans; these membranes were dilated and eventually destroyed.[445] With cucumbers, it was found that cobalt promotes hypocotyl elongation of seedlings by inhibiting ethylene production.[446] In Sweden, it has been reported that in organic soils the ability of plants to utilize added cobalt may be predicted from pH measurements.[447]

Cobalt additions have been found to decrease the incidence of common scab on potatoes,[322] and to increase the thiamine, nicotinic acid, and ascorbic acid in lupine leaves.[448] The chlorophyll compound obtained with bivalent cobalt corresponded to the molecular ratio chlorophyll:metal = 3:1.[449]

References

1. Bertrand, D. and Mokragnatz, M. (1930). *Bull. Soc. Chim.* **47**, 326–31.
2. Young, R. S. (1960). "Cobalt", A.C.S. Monograph 149. Reinhold, New York.
3. Aleksiev, A., Krusteva, E., and Kumanov, S. (1964). *Zhivotnovudni Nauki* **1**, 45–53; (1965). *C.A.* **62**, 9465.
4. Grujic-Injac, B., Micic, D., and Stojanovic, N. (1963). *Hem. Društ., Beograd.* **28**, 439–45; (1965). *C.A.* **63**, 15221.
5. Tsingovatov, V. A. and Beletkova, L. S. (1964). Mikroelem. Biosfere Ikh Primen. Sel'. Khoz. Med. Sib. Dal'nego Vostoka, Dokl. Sib. Konf. 2nd, 121–4; (1969). *C.A.* **70**, 86666.
6. Drobiz, F. D. and Kadochnikova, A. A. (1970). *Okhr. Prir. Urale* **7**, 163–6; (1971). *C.A.* **75**, 34578.
7. Prazdnikova, R. V. and Peterburgskii, A. V. (1971). *Dokl. TSKHA* No. 172, 116–20; (1972). *C.A.* **77**, 4037.
8. Staskiewicz, G., Wiercinski, J., and Fidecka, H. (1972). *Ann. Univ. Mariae Curie-Sklodowska, Sect. DD* **27**, 151–64; (1974). *C.A.* **81**, 168488.
9. Koter, M. and Krauze, A. (1968). *Rocz. Glebozn.* **19**, 121–33; (1969). *C.A.* **71**, 2617.
10. Yagodin, B. A. and Troitskaya, G. N. (1974). Biol. Rol Mikroelem. Ikh Primen. Sel'. Khoz. Med., 329–38; (1975). *C.A.* **82**, 121616.
11. Mocsy, J. and Tolgyesi, G. (1960). *Magy. Allatorv. Lap.* **15**, 44–7; (1960). *C.A.* **54**, 16688.
12. Reddy, K. G. and Mehta, B. V. (1961). *Indian J. Agric. Sci.* **31**, 108–12; (1962). *C.A.* **56**, 9178.
13. Garcilaso, S. G. (1961). *Archos Venez. Nutr.* **11**, 109–19; (1962). *C.A.* **56**, 6419.
14. Obukhova, Z. D., Pirogova, N. V., Dorozhlina, A. F., and Shpolyanskaya, E. I. (1962). Mikroelementy v Zhivotnovod. i Rastenievod., Akad. Nauk Kirg. S.S.R., 71–106; (1963). *C.A.* **58**, 11733.
15. Petrova, R. and Radenkov, S. (1969). *Gradinar. Lozar. Nauka* **6**, 115–17; (1969). *C.A.* **71**, 80311.
16. Platash, I. T., Deryugina, L. I., and Artemchenko, V. S. (1972). *Farm. Zh.* **27**, 64–5; (1972). *C.A.* **77**, 4233.

17. Beeson, K. C., Gray, L., and Adams, M. B. (1947). *J. Am. Soc. Agron.* **39**, 356–62.
18. Zizuma, A. and Strik, Z. (1957). *Trudy Inst. Eks. Med., Akad. Nauk Latv. S.S.R.* **14**, 197–9; (1959). *C.A.* **53**, 1483.
19. Openlender, I. V. (1969). *Mikroelem. Zhivotnovod. Rastenievod.* No. 8, 64–70; (1971). *C.A.* **74**, 31210.
20. Oparin, I. A. (1968). *Vop. Pitan.* **27**, 72–3; (1969). *C.A.* **70**, 56467.
21. Rudin, V. D. (1962). *Trudy Stavropol. Sel'.-khoz. Inst.* 21–7; (1964). *C.A.* **60**, 4723.
22. Patel, B. M., Vaidya, M. B., and Mistry, V. V. (1966). *J. Vet. Sci. Anim. Husb.* **36**, 130–4.
23. Sadovskaya, E. N., Kurinnyi, V. D., and Mukhin, V. P. (1973). *Dokl. TSKHA* **193**, 37–40; (1974). *C.A.* **81**, 76876.
24. Obukhova, Z. D., Dorozhlina, A. F., Pirogova, N. V., and Saenkova, R. E. (1968). Mikroelem. Zhivotnovod. Rastenievod., 96–124; (1970). *C.A.* **72**, 99362.
25. Skoropostizhnaya, A. S. (1957). *Vop. Pitan.* **16**, 59–62; (1957). *C.A.* **51**, 1599.
26. Morozov, V. I. and Pimenov, P. K. (1972). *Trudy Ul'yanovsk. Sel'.-khoz. Inst.* **17**, 63–8; (1973). *C.A.* **79**, 102722.
27. Hurwitz, C. and Beeson, K. C. (1944). *Food Res.* **9**, 348–57.
28. Mitchell, J. H. (1951). S. Carolina Agric. Exp. Stn Bull. 391.
29. Arthur, D. Motzok, I., and Branion, H. D. (1953). *Can. J. Agric. Sci.* **33**, 1–15.
30. Grinkevich, N. I., Gribovskaya, I. F., Shandova, A. N., and Dinevich, L. S. (1971). *Biol. Nauk* **14**, 88–91; (1971). *C.A.* **75**, 31423.
31. Rush, V. A. and Ligunova, V. V. (1969). *Vop. Pitan.* **28**, 52–5; (1969). *C.A.* **71**, 809.
32. Kandatsu, M. and Mori, B. (1956). *Nippon Nogei-Kagaku Kaishi* **30**, 100–5; (1957). *C.A.* **51**, 9969.
33. Suciu, T. and Ivanof, L. (1963). *Acad. Rep. Populare Romine, Filiala Cluj, Studii Cercetari Agron.* **14**, 79–81; (1965). *C.A.* **63**, 13982.
34. Podusowska, I., Krawczyk, K., and Ombach, A. (1969). *Rocz. Nauk. Roln., Ser. B* **91**, 573–83; (1970). *C.A.* **73**, 73897.
35. Nicolls, K. D. and Honeysett, J. L. (1964). *Aust. J. Agric. Res.* **15**, 368–76.
36. Askew, H. O. and Dixon, J. K. (1937). *New Zealand J. Sci. Technol.* **18**, 688–93.
37. Leonov, V. A., Terent'eva, M. V., and Gorski, N. A. (1960). *Vest. Akad. Navuk Belaruss. S.S.R., Ser. Biyal Navuk* No. 3, 47–55; (1962). *C.A.* **56**, 15843.
38. Khakimova, V. K., Galkina, N. V., and Taubaev, T. T. (1971). *Khim. Sel'. Khoz.* **9**, 787–8; (1972). *C.A.* **76**, 98442.
39. Dobrolyubskii, O. K., Ryzha, V. K., Fedorenko, I. V., and Zhivitskaya, L. I. (1961). Rol Mikroelementov v Sel'. Khoz. Tr. 2-go (Vtorogo) Mezhvuz, Soveshch. po Mikroelementam, 114–23; (1962). *C.A.* **57**, 8929.
40. Chesev, K. S. and Gerasimova, L. K. (1968). *Vop. Pitan.* **27**, 69–72; (1969). *C.A.* **70**, 56466.
41. Kruglyakov, G. N. (1972). *Izv. Vyssh. Ucheb. Zaved. Pishch. Tekhnol.* **4**, 177–9; (1973). *C.A.* **78**, 41780.
42. Pickett, E. E. (1955). Missouri Agric. Exp. Stn Res. Bull. 594.
43. Mlodecki, H., Lasota, W., and Tersa, S. (1965). *Farmaceuta Pol.* **21**, 337–43; (1965). *C.A.* **63**, 13941.
44. Bowstead, J. E., Sackville, J. P., and Sinclair, R. D. (1942). *Sci. Agric.* **22**, 314–25.
45. Twist, J. O., Morris, J. G., and Gartner, R. J. W. (1965). *Aust. J. Sci.* **28**, 125–6.
46. Kamynina, L. M. (1964). *Dokl. Rossinsk. Sel'.-khoz. Akad.* No. 99, 319–21, (1966). *C.A.* **64**, 2702.

47. Yur'eva, V. I. (1962). *Trudy̆ Troitsk. Vet. Inst.* **8**, 20–5; (1964). *C.A.* **60**, 11046.
48. Dube, J. N. (1964). *J. Indian Soc. Soil Sci.* **12**, 381–5; (1966). *C.A.* **64**, 5527.
49. Meleshko, K. V. (1956). *Vop. Pitan.* **15**, 43–7; (1957). *C.A.* **51**, 6037.
50. Vinogradova, Z. A. and Kovaljskii, V. V. (1962). *Dokl. Akad. Nauk S.S.S.R.* **147**, 1458–60; (1963). *C.A.* **58**, 14441.
51. Goto, T. (1954–55). *Eiyo to Shokuryo* **7**, 102–3; (1959). *C.A.* **53**, 7448.
52. Coppenet, M., More, E., LeCorre, L., and Le Mao, M. (1972). *Annls Agron.* **23**, 165–96; (1972). *C.A.* **77**, 113044.
53. Vinogradov, A. P. (1953). "The Elementary Chemical Composition of Marine Organisms". Sears Foundation for Marine Research, New Haven.
54. Lakanen, E. (1969). *Annls Agric. Fenn.* **8**, 20–9; (1969). *C.A.* **71**, 122584.
55. Micevska, M., Jordanovska, V., and Tosev, D. (1974). *Godisen Zb. Prir.-Mat. Fak. Univ. Skopje, Mat. Fiz. Hem.* **24**, 89–91; (1975). *C.A.* **82**, 135986.
56. Gumenyuk, V. D., Burtsev, V. Ya., P'yanova, N. M., and Ionova, L. Yu. (1974). *Visn. Sil's'kogospod. Nauki* **6**, 55–8; (1974). *C.A.* **81**, 118801.
57. Bergman, A. G. and Khoroshkina, L. I. (1969). *Sb. Nauch. Tr. Donskoi Sel'.-khoz. Inst.* **6**, 103–9; (1972). *C.A.* **77**, 60304.
58. Jungerman, K. (1960). *Landw. Forsch.* **13**, 153–67; (1960). *C.A.* **54**, 25464.
59. Volostnov, G. A. and Lachko, O. A. (1971). *Mikroelem. Zhivotnovod. Rastenievod.* No. 10, 84–8; (1973). *C.A.* **79**, 77185.
60. Gritsova, G. I. and Toikka, M. A. (1964). *Uchen. Zap. Petrozavodsk. Gos. Univ.* **12**, 28–31; (1966). *C.A.* **65**, 6237.
61. Gyul'akhmedov, A. N. and Gadzhiev, F. M. (1968). *Mater. Respub. Konf. Probl. "Mikroelem. Med. Zhivotnovod"*, 1st, 11–13; (1970). *C.A.* **73**, 55056.
62. Obukhova, Z. D. and Markelova, S. V. (1966). *Mikroelem. Zhivotnovod. Rastenievod. Akad. Nauk Kirg. S.S.R.*, 127–36; (1967). *C.A.* **66**, 75117.
63. Boenig, G. (1961). *Phosphorsäure* **21**, 298–301; (1962). *C.A.* **56**, 15906.
64. Shlyakman, M. Ya., Krasutskaya, N. L., and Yavorskaya, T. S. (1965). *Mikroelementy v Zhivotnovod. i Med.*, Akad. Nauk Ukr. S.S.R., Resp. Mezhvedomstv. Sb., 19–24; (1966). *C.A.* **64**, 13295.
65. Ch'en, C. T., Wei, C. T., Yeh, C. P., and Cheng, J. M. (1966). *Chung Kuo Nung Yeh Hua Hsueh Hui Chih* **4**, 19–24; (1967). *C.A.* **66**, 75319.
66. Bottini, E., Sapetti, C., and di Lavriano, E. M. (1959). *Annali Sper. Agra.* **13**, 499–522; (1959). *C.A.* **53**, 22299.
67. Coppenet, M., Le Corre, L., Guillou, M., and More, E. (1970). *C.R. Hebd. Séances Acad. Agric. Fr.* **56**, 923–37; (1971). C.A. **74**, 140215.
68. Iyer, J. G. and Satyanaryan, Y. (1958). *Current Sci.* **27**, 220–1; (1959). *C.A.* **53**, 4434.
69. Hayakawa, T. (1962). *Nat. Inst. Anim. Hth Quart.* **2**, 172–81; (1963). *C.A.* **58**, 10511.
70. Mehta, B. V., Reddy, G. R., Nair, G. K., Ghandi, S. C., Neelkantan, V., and Reddy, K. G. (1964). *J. Indian Soc. Sci.* **12**, 329–42; (1966). *C.A.* **64**, 7311.
71. Long, M. I. E. and Frederiksen, S. (1970). *Z. Pfl-Ernähr. Düng. Bodenk.* **126**, 238–44; (1971). *C.A.* **74**, 86945.
72. Steger, H., Franz, H., and Pueschel, F. (1966). *Arch. Tierernähr.* **16**, 215–28; (1966). *C.A.* **65**, 2949.
73. Kazaryan, E. S., Asratyan, G. S., and Stepanyan, M. S. (1965). *Soobshch. Lab. Agrokhim.*, Akad. Nauk Arm. S.S.R. No. 6, 20–30; (1966). *C.A.* **65**, 6236.
74. Obukhova, Z. D. and Markelova, S. V. (1968). *Mikroelem. Zhivotnovod. Rastenievod.*, 125–31; (1970). *C.A.* **72**, 75674.

75. Belovodskii, V. L. (1972). *Uchen. Zap., Dal'nevost. Gos. Univ.* **57**, 49–55; (1974). *C.A.* **80**, 69379.
76. Kurilyuk, T. T. (1964). Biokhim. Osobennosti Rast. Yakutii, Akad. Nauk S.S.S.R., Yakutskii Filial Sibirsk. Otdel., Inst. Biol., 107–17; (1965). *C.A.* **62**, 12153.
77. Kozyreva, G. F. and Rish, M. A. (1965). Mikroelementy v Sel'. Khoz. Akad. Nauk Uzbek. S.S.R., Otdel. Khim-Tekhnol. i Biol. Nauk, 227–31; (1966). *C.A.* **64**, 11804.
78. Hayakawa, T. (1961). *Igaku To Seibutsugaku* **61**, 85–7; (1964). *C.A.* **60**, 14829.
79. Jungerman, K. (1961). *Landw. Forsch.* **14**, 131–5; (1962). *C.A.* **56**, 13312.
80. Coppenet, M. and Calvez, J. (1967). *C.R. Hebd. Séances Acad. Agric. Fr.* **53**, 939–47; (1968). *C.A.* **68**, 68042.
81. Makmudov, Kh. Kh. (1972). *Vest. Sel'.-khoz. Nauki* **15**, 119–22; (1973). *C.A.* **78**, 83234.
82. Kurilyuk, T. T. (1967). Mikroelem. Biosfere Ikh Primen. Sel'.-khoz. Med. Sib. Dal'nevost., 115–20; (1969). *C.A.* **70**, 2764.
83. Kudashev, A. K. (1972). *Khim. Sel'. Khoz.* **10**, 701–2; (1973). *C.A.* **78**, 28275.
84. Obukhova, Z. D. and Dorozhkina, A. P. (1971). *Mikroelem. Zhivotnovod. Rastenievod.* No. 10, 94–8; (1973). *C.A.* **79**, 77184.
85. Kummer, H., Von Polheim, P., and Scholl, W. (1973). *Landw. Forsch., Sonderh.* **28**, 215–27; (1974). *C.A.* **80**, 81417.
86. Hasler, A. and Zuber, R. (1955). *Schweiz. Landw. Monatsh.* **33**, 192–202; (1959). *C.A.* **53**, 606.
87. Nehring, K. and Borchmann, W. (1959). *Z. Landw. Versuchs-u. Untersuchungsw.* **5**, 556–70; (1961). *C.A.* **55**, 6724.
88. Gericke, S. (1962). *Phosphorsäure* **22**, 48–60; (1963). *C.A.* **58**, 2813.
89. Kiryukhin, R. A. (1959). *Konevodstvo* No. 7, 31–2; (1963). *C.A.* **59**, 4318.
90. Wind, J. (1957). *Landbouwk. Tijdschr.* **69**, 608–18; (1960). *C.A.* **54**, 20005.
91. Kazaryan, E. S. and Airuni, G. A. (1967). *Izv. Sel'.-khoz. Nauk, Min. Sel. Khoz. Arm. S.S.R.* **10**, 65–72; (1968). *C.A.* **68**, 28789.
92. Ben-Utyaeva, G. S. (1965). Mikroelementy v Sel'. Khoz. Sb., 318–21; (1966). *C.A.* **64**, 10352.
93. Kotelyanskaya, L. I. (1963). *Vop. Pitan.* **22**, 71–2; (1964). *C.A.* **60**, 7359.
94. Panin, M. S. and Panina, R. I. (1971). *Biol. Nauki* **14**, 69–75; (1972). *C.A.* **76**, 33223.
95. Kazaryan, E. S. (1969). *Gorn. Luga, Ikh Uluchshenie Ispol'z*, 87–94; (1970). *C.A.* **72**, 51788.
96. Staskiewicz, G., Fidecka, H., Wiercinski, J., and Zimowska, K. (1969). *Ann. Univ. Mariae Curie-Sklodowska, Sect. DD* **24**, 159–64; (1971). *C.A.* **75**, 115847.
97. Staskiewicz, G., Wiercinski, J., Zimowska, K., and Fidecka, H. (1971). *Ann. Univ. Mariae Curie-Sklodowska, Sect. DD* **26**, 109–13; (1973). *C.A.* **78**, 94888.
98. Egorova, T. K. (1959). *Udobr. Urozh.* **4**, 32–5; (1960). *C.A.* **54**, 25458.
99. Kozyreva, G. F. (1965). Mikroelementy v Sel'. Khoz. Akad. Nauk Uzbek. S.S.R., Otdel. Khim.-Tekhnol. i Biol. Nauk, 313–17; (1966). *C.A.* **64**, 13335.
100. Rish, M. A., Ben-Utyaeva, G. S., Kozyreva, G. F., and Priev, Ya. M. (1963). *Trudy Inst. Karakulevodstva, Min. Sel'. Khoz. Uzbek. S.S.R.* **13**, 379–95; (1965). *C.A.* **63**, 2343.
101. Marano, B. and Rainone, S. (1959). *Ricerca Scient.* **29**, 133; (1959). *C.A.* **53**, 19217.
102. Lambert, T. L. and Blincoe, C. (1971). *J. Sci. Food Agric.* **22**, 8–9.

103. Warren, H. V. and Delavault, R. (1957). *Trans. R. Soc. Canada, Third Ser. Section IV* **51**, 33–7.
104. Young, R. S. (1974). *Commonw. For. Rev.* **53**, 49–51.
105. Guha, M. M. and Mitchell, R. L. (1965). *Pl. Soil* **23**, 323–38.
106. Gyul'akhmedov, A. N., Kuliev, Sh. M., and Gyandzhemekhr, A. V. (1974). Trudȳ Azerb. Fil. Vses. O-va. Pochvovedov, 48–51; (1974). *C.A.* **84**, 72936.
107. Kubota, J., Lazar, V. A., and Beeson, K. C. (1960). *Soil Sci. Soc. Am. Proc.* **24**, 527–8.
108. Beaton, J. D., Brown, G., Speer, R. C., MacRae, I., McGhee, W. P. T., Moss, A., and Kosick, R. (1965). *Soil Sci. Soc. Am. Proc.* **29**, 299–302.
109. Vasil'kov, V. V. and Golodushko, B. Z. (1971). *Dokl. Akad. Nauk Beloruss. S.S.R.* **15**, 944–6; (1972). *C.A.* **76**, 58259.
110. Lazar, V. A. and Beeson, K. C. (1956). *J. Agric. Food Chem.* **4**, 439–44.
111. Alban, L. A. and Kubota, J. (1960). *Soil Sci. Soc. Am. Proc.* **24**, 183–5.
112. Thomas, W. A. (1975). *Forest Sci.* **21**, 222–6.
113. Yamagata, N. and Murakami, Y. (1958). *Nature, Lond.* **181**, 1808.
114. Pickett, E. E. (1960). Missouri Agric. Exp. Stn Res. Bull. 724.
115. Price, N. O. and Hardison, W. A. (1963). Virginia Agric. Exp. Stn Bull. 165.
116. Price, N. O. and Huber, J. T. (1964). Virginia Agric. Exp. Stn Bull. 177.
117. Kubota, J. (1964). *Soil Sci. Soc. Am. Proc.* **28**, 246–51.
118. Ignat'ev, V. N. and Levin, M. M. (1968). *Trudȳ Mold. Nauch.-Issled. Inst. Zhivotnovod. Vet.* **4**, 170–2; (1972). *C.A.* **77**, 138538.
119. Pereira, J. A. A., Da Silva, D. J., and Braga, J. M. (1971). *Experientia* **12**, 155–88; (1972). *C.A.* **76**, 125761.
120. Rampilova, M. A., Petrovich, P. I., Belokurova, E. I., Emedeeva, N. I., and Kharitonov, Yu. D. (1974). *Mikroelem. Sib.* **9**, 114–18; (1975). *C.A.* **83**, 77227.
121. Kossila, V. (1976). *Karjanticote* **59**, 11; (1976). *C.A.* **85**, 107722.
122. van der Merwe, F. J. (1959). *S. African J. Agric. Sci.* **2**, 141–63.
123. Correa, R. (1957). *Arq. Inst. Biol.* **24**, 199–227; (1960). *C.A.* **54**, 21359.
124. Odynets, R. N. (1966). Mikroelem. Sel'. Khoz. Med. Dokl. Vses. Soveshch. Mikroelem. 5th, 507–13; (1970). *C.A.* **73**, 1321.
125. Odynets, R. N. and Aituganov, M. D. (1969). *Izv. Akad. Nauk Kirg. S.S.R.* No. 4, 48–58; (1970). *C.A.* **72**, 118938.
126. Golosov, I. M. and Zostautas, A. (1971). *Khim. Sel'. Khoz.* **9**, 276–7; (1971). *C.A.* **75**, 46148.
127. Lapin, L. N. (1968). Mater.. Respub. Konf. Probl. "Mikroelem. Med. Zhivotnovod." 1st, 137–8; (1971). *C.A.* **74**, 29481.
128. Chubinskaya, A. A. (1962). Mikroelementy v Zhivotnovod. Min. Sel'. Khoz. S.S.R. Vses. Akad. Sel'.-khoz. Nauk, Otdel. Zhivotnovod. 107–13; (1963). *C.A.* **58**, 11738.
129. Dzinic, M. et. al. (1960). *Veterinaria* **9**, 679–90; (1964). *C.A.* **60**, 7141.
130. Chodan, J. (1962). *Rocz. Nauk Roln., Ser. F* **75**, 545–62; (1963). *C.A.* **59**, 15889.
131. Uzilovskaya, P. Sh., Uspenskaya, M. V., and Ivshina, V. I. (1962). *Trudȳ Nauch.-Issled. Inst. Zhivotnovod. Uzbek. Akad. Sel'.-khoz. Nauk* No. 7, 27–31; (1963). *C.A.* **59**, 5559.
132. Korovin, N. K. (1964). Mikroelem. Biosfere Ikh Primen. Sel'. Khoz. Med. Sib. Dal'nego Vostoka, Dokl. Sib. Konf. 2nd, 464–6; (1969). *C.A.* **70**, 65709.
133. Gustun, M. I. (1965). *Khim. Sel'. Khoz.* **3**, 67–8; (1966). *C.A.* **64**, 7311.
134. Gustun, M. I. (1965). *Trudȳ Vses. Nauch.-Issled. Inst. Fiziol. Biokhim. Sel'.-khoz. Zhivotn.* **2**, 371–85; (1967). *C.A.* **66**, 64662.

135. Muzaleva, L. D. and Pershina, E. F. (1965). *Uchen. Zap. Petrozavodsk. Gos. Univ.* **13**, 21–8; (1966). *C.A.* **65**, 19254.
136. Mecheva, K. T. (1966). *Uchen. Zap. Kabardino-Balkar. Gos. Univ.* No. 28, 317–22; (1969). *C.A.* **71**, 37588.
137. Kotysheva, N. G. and Koverga, L. V. (1967). *Mikroelem. Zhivotnovod. Rastenievod. Akad. Nauk Kirg. S.S.R.* No. 6, 59–63; (1968). *C.A.* **69**, 94116.
138. Chupakhina, K. G. (1968). *Khim. Sel'. Khoz.* **6**, 500–3; (1969). *C.A.* **70**, 2580.
139. Rudin, V. D., Shcherbakova, S. S., Churilova, M. I., and Mishustina, A. T. (1969). *Trudy Stavropol. Sel'.-khoz. Inst.* No. 32, 157–60; (1970). *C.A.* **73**, 87073.
140. Abou-Hussein, E. R. M., Raafat, M. A., Abou-Raya, A. K., and Shalaby, A. S. (1970). *U.A.R. J. Anim. Prod.* **10**, 245–54; (1973). *C.A.* **78**, 28186.
141. Klimakhin, N. A. (1970). *Vest. Sel'.-khoz. Nauki* **13**, 85–8; (1970). *C.A.* **73**, 75841.
142. Iyer, I. G. (1959). *J. Biol. Sci.* **1**, 101–9.
143. Anderson, A. J. (1970). *J. Aust. Inst. Agric. Sci.* **36**, 15–29.
144. Stewart, J., *et al.* (1955). *Vet. Record* **67**, 755–6.
145. Kabata, A. and Beeson, K. C. (1960). *Rocz. Nauk Roln., Ser. A.* **83**, 277–89; (1961). *C.A.* **55**, 23890.
146. Taubel, N. (1961). *Landw. Forsch.* **14**, 188–93; (1962). *C.A.* **57**, 1306.
147. Hayakawa, T. and Takayama, T. (1961). *Igaku to Seibutsugaku* **60**, 49–51; (1964). *C.A.* **60**, 2289.
148. Terent'eva, M. V. and Lobach, T. Ya. (1963). *Vest. Akad. Navuk Belaruss. S.S.R., Ser. Biyal. Navuk* No. 3, 59–63; (1963). *C.A.* **60**, 9857.
149. Kuehn, H. (1962). *Sonderh. Z. "Landw. Forsch."* No. 16, 112–21; (1963). *C.A.* **58**, 10684.
150. Potakhina, L. N. (1965). *Uchen. Zap. Petrozavodsk. Gos. Univ.* **13**, 18–20; (1966). *C.A.* **65**, 19260.
151. Furlan, J. and Stupar, J. (1967). *Zemlj. Biljka* **16**, 691–6; (1968). *C.A.* **68**, 113629.
152. Yagodin, B. A. and Ovcharenko, G. A. (1969). *Agrokhimiya* No. 3, 101–4; (1969). *C.A.* **71**, 2590.
153. Mitchell, R. L. (1972). *Atti Simp. Int. Agrochim.* **9**, 521–32.
154. Kenesarina, N. A. (1972). *Izv. Akad. Nauk Kaz. S.S.R. Ser. Biol.* **10**, 31–5; (1973). *C.A.* **78**, 96561.
155. Kabata-Pendias, A. (1973). *Rocz. Glebozn.* **24**, 273–87; (1974). *C.A.* **80**, 94605.
156. Bambergs, K. (1955). Mikroelementy v Sel'. Khoz. i Med. Akad. Nauk Latv. S.S.R., Otdel. Biol. Nauk, Tr. Vses. Sovesch. Riga, 67–80; (1959). *C.A.* **53**, 10624.
157. Anke, M. (1961). *Z. Acker-u. PflBau* **112**, 113–40; (1961). *C.A.* **55**, 19088.
158. Hayakawa, T. (1961). *Igaku to Seibutsugaku* **60**, 151–4; (1964). *C.A.* **60**, 2045.
159. Andrews, E. D. (1966). *New Zealand J. Agric. Res.* **9**, 829–38.
160. Kocialkowski, Z., Czekalski, A., and Baluk, S. (1967). *Rocz. Nauk Roln., Ser. A.* **93**, 155–76; (1968). *C.A.* **68**, 21218.
161. Makarov, V. A. (1971). *Zap. Leningrad. Sel'.-khoz. Inst.* **160**, 23–6; (1972). *C.A.* **76**, 58252.
162. Travnikova, L. S. (1964). *Vest. Mosk. Univ., Ser. VI., Biol. Pochvoved.* **19**, 74–80; (1964). *C.A.* **61**, 8844.
163. Gulyakin, I. V. and Yudintseva, E. V. (1958). *Dokl. Akad. Nauk S.S.S.R.* **123**, 368–70; (1959). *C.A.* **53**, 4437.
164. Gulyakin, I. V. and Yudintseva, E. V. (1960). *Izv. Timiryazev. Sel'.-khoz. Akad.* No. 5, 114–22. (1961). *C.A.* **55**, 15800.

165. Yudintseva, E. V. (1961). Rol Mikroelementov v Sel'. Khoz. Tr. 2-go (Vtorogo) Mezhvuz. Soveshch. po Mikroelementam, 177–85; (1962). *C.A.* **57**, 8913.
166. Makhonina, G. I., Molehanova, I. V., Subbotina, E. N., Timofeev-Resovskii, N. V., Titlyanova, A. A., Tyurukanov, A. N., and Chebotina, M. Ya. (1965). *Trudȳ Inst. Biol., Akad. Nauk S.S.R., Ural'sk. Filial* No. 45, 121–5; (1966). *C.A.* **65**, 4587.
167. Terent'eva, M. V. and Dorozhkina, L. N. (1967). *Agrokhimiya* No. 2, 67–71; (1967). *C.A.* **67**, 746.
168. Mirzaeva, K. Kh. (1961). *Dokl. Akad. Nauk Uzbek. S.S.R.* No. 2, 24–6; (1964). *C.A.* **61**, 960.
169. Panteleeva, E. I. (1956). *Zap. Leningrad. Sel'.-khoz. Inst.* No. 11, 221–4; (1959). *C.A.* **53**, 2373.
170. Veretka, M. S. and Kobets, N. M. (1966). *Visn. Sil's'kogospod. Nauki* **9**, 80–1; (1967). *C.A.* **66**, 64663.
171. Nartov, V. I. (1962). *Trudȳ Stavropol. Sel'.-khoz. Inst.* No. 23, 96–8; (1968). *C.A.* **69**, 35037.
172. Carles, J., Cabrol., P., and Magny, J. (1961). *Bull. Soc. Chim. Biol.* **43**, 1111–20; (1962). *C.A.* **56**, 2706.
173. Toikka, M. A. (1964). *Uchen. Zap. Petrozavodsk. Gos. Univ.* **12**, 57–63; (1966). *C.A.* **65**, 7952.
174. Minasyan, S. M. and Martirosyan, M. Ya. (1966). *Dokl. Akad. Nauk Arm. S.S.R.* **42**, 166–9; (1966). *C.A.* **65**, 15788.
175. Uesaka, S., Kawashima, R., Hashimoto, Y., and Kamada, Y. (1959). *Kyoto Daigaku Shokuryo Kagaku Kenkyusho Hokoku* No. 22, 17–35; (1960). *C.A.* **54**, 1675.
176. Dobrolyubskii, O. K. and Zhivitskaya, L. I. (1960). *Nauch. Dok. Vȳssh. Shk., Biol. Nauki* No. 4, 186–9; (1961). *C.A.* **55**, 10601.
177. Liwski, S. (1960). *Zesz. Probl. Postep. Nauk Roln.* No. 25, 197–240; (1964). *C.A.* **61**, 4706.
178. Loper, G. M. and Smith, D. (1961). Wisconsin Univ. Agric. Exp. Stn Res. Rept. 8.
179. Hayakawa, T. (1961). *Igaku to Seibutsugaku* **60**, 165–7; (1964). *C.A.* **60**, 2045.
180. Kulikov, N. V. (1961). *Pochvovedenie* No. 4, 79–81; (1961). *C.A.* **55**, 20278.
181. Panova, S. V. and Skobeleva, N. N. (1965). Mikroelementy v Zhivotnovod. i Med. Akad. Nauk Ukr. S.S.R., Resp. Mezhvedomstv. Sb., 11–18; (1966). *C.A.* **64**, 14912.
182. Kee, N. S. and Bloomfield, C. (1961). *Geochim. Cosmochim. Acta* **24**, 206–25.
183. Piotrowska, M. (1974). *Rocz. Glebozn.* **25**, 157–67; (1975). *C.A.* **82**, 124035.
184. Lounamaa, J. (1956). *Ann. Botan. Soc. Zool. Botan. Fennicae "Vanamo"* No. 4; (1959). *C.A.* **53**, 5418.
185. Agafanova, A. F. (1955). Mikroelementy v Sel'. Khoz. i Med., Akad. Nauk Latv. S.S.R., Otdel. Biol. Nauk, Tr. Vses. Soveshchan. Riga, 213–19; (1959). *C.A.* **53**, 11539.
186. Reith, J. W. S. and Mitchell, R. L. (1962). Plant Anal. Fertilizer Probl. 4th, Brussels, 241–54.
187. Modor, V. and Tolgyesi, G. (1964). *Kísérl. Közl. B. Alláttényesz.* **57**, 59–66; (1966). *C.A.* **65**, 2949.
188. Nicolls, K. D. and Honeysett, J. L. (1964). *Aust. J. Agric. Res.* **15**, 609–24.
189. Gille, G. L. and Graham, E. R. (1971). *Soil Sci. Soc. Am. Proc.* **35**, 414–16.

190. Lukashev, K. I. and Petukhova, N. N. (1962). *Dokl. Akad. Nauk Beloruss. S.S.R.*
     6, 448–52; (1962). *C.A.* **57**, 14204.
191. Biswas, N. R. D. (1964). *J. Indian Soc. Soil Sci.* **12**, 375–9.
192. Travnikova, L. S. (1963). *Nauch. Dokl. Vyssh. Shk., Biol. Nauki* No. 2, 192–5;
     (1966). *C.A.* **65**, 2954.
193. Adams, S. N., Honeysett, J. L., Tiller, K. G., and Norrish, K. (1969). *Aust. J. Soil
     Res.* **7**, 29–42.
194. Filipovic, Z., Stankovic, B., Dusic, Z., and Pisteljic, V. (1959). *Arh. Poljopr.
     Nauke Teh* **12**, 63–8; (1960). *C.A.* **54**, 10075.
195. Young, R. S. (1935). Cornell Univ. Agric. Exp. Stn Memoir 174.
196. Young, R. S. (1951). *S. African Ind. Chem.* **5**, 196–7.
197. Young, R. S. (1953). *Nutr. Obs.* **14**, 1–3.
198. Young, R. S. (1956). *Sci. Prog.* **44**, 16–37.
199. Dufrenoy, J. (1959). *Revue Path. Gén. Physiol. Clin.* **59**, 257–63; (1959). *C.A.* **53**,
     11735.
200. West, S. H. and Harris, H. C. (1965). *Soil Crop Sci. Soc. Fla. Proc.* **25**,
     83–95.
201. Sanchelli, V. (1969). "Trace Elements in Agriculture". Van Nostrand–Reinhold,
     New York.
202. Yagodin, B. A. (1970). "Cobalt in the Life of Plants". Nauka, Moscow.
203. Koval'skii, V. V., Raetskaya, Yu. I., and Gracheva, T. I. (1971). "Trace Elements
     in Plants and Foods". Kolos, Moscow.
204. Lisk, D. J. (1972). *Adv. Agron.* **24**, 267–325.
205. Donald, C. M. and Prescott, J. A. (1975). *In* "Trace Elements in Soil–Plant–
     Animal Systems" (D. J. D. Nicholas and A. R. Egan, eds) 7–37. Academic Press,
     New York and London.
206. Loneragan, J. F. (1975). *In* "Trace Elements in Soil–Plant–Animal Systems"
     (D. J. D. Nicholas and A. R. Egan, eds) 109–34. Academic Press, New York
     and London.
207. Nicholas, D. J. D. (1975). *In* "Trace Elements in Soil–Plant–Animal Systems"
     (D. J. D. Nicholas and A. R. Egan, eds) 181–98. Academic Press, New York and
     London.
208. Reuter, D. J. (1975). *In* "Trace Elements in Soil–Plant–Animal Systems" 291–
     324. Academic Press, New York and London.
209. Petrunina, N. S. (1974). *Trudy Biogeokhim. Lab., Akad. Nauk S.S.S.R.* **13**, 57–
     117; (1974). *C.A.* **81**, 103701.
210. Nicholas, D. J. D. and Egan, A. R. (1975). "Trace Elements in Soil–Plant–
     Animal Systems". Academic Press, New York and London.
211. Rozenberg, R. E. (1958). Osnown. Rezul'taty Nauch.-Issled. Raboty Nauch.-
     Issled. Inst. Melior i Vodn. Khoz. Akad. Sel'.-khoz. Nauk Beloruss. S.S.R.,
     Minsk, Sbornik, 212–19; (1962). *C.A.* **56**, 5142.
212. Busse, M. (1959). *Planta* **53**, 25–44; (1959). *C.A.* **53**, 14239.
213. Semina, R. M. (1967). *Nauch Dokl. Vyssh., Shk. Biol. Nauki* **3**, 80–3; (1967). *C.A.*
     67, 10806.
214. Agapov, A. I. and Simakov, V. N. (1968). Vop. Kornevogo Pilan. Rast., 99–116;
     (1969). *C.A.* **71**, 90392.
215. Ivanou, M. P., Ivakhenka, N. M., Lobach, T. Ya., Vashkevich, L. F., Svirnouski,
     L. Ya., and Lyakhovich, S. R. (1975). *Vest. Akad. Navuk Beloruss. S.S.R.*, *Ser.
     Sel'skagaspad. Navuk* No. 2, 75–8; (1975). *C.A.* **83**, 95492.
216. Jungerman, K. (1960). *Landw. Forsch.* **13**, 153–67; (1960). *C.A.* **54**, 25464.
217. Minina, E. I. (1973). *Trudy Gor'k. v. Sel'.-khoz. Inst.* **55**, 175–80; (1975). *C.A.* **82**,
     42270.

218. Asmus, F. (1961). *Dt. Landwirt.* No. 2, 72–4; (1961). *C.A.* **55**, 20288.
219. Kraznopertsev, N. G. (1966). *Uchen. Zap. Novgorod. Golovnoi Gos. Pedagog. Inst.* **5**, 28–42; (1969). *C.A.* **70**, 36698.
220. Elagin, I. N. (1970). *Dokl. Vses. Akad. Sel'.-khoz. Nauk* No. 7, 22–3; (1970). *C.A.* **73**, 97986.
221. Minina, E. I. (1973). *Trudÿ Gor'kov. Sel'-khoz. Inst.* **55**, 187–90; (1975). *C.A.* **82**, 42272.
222. Lambin, A. (1959). *Sel'. Khoz. Sibiri* No. 1, 56–60; (1961). *C.A.* **55**, 1987.
223. Efimov, M. V. (1960). Mikroelementy v Pochvakh i Organizmakh Vost. Siberi i Dal'nego Vostoka i ikh Rol v Zhizni Rast. Zhivotn. i Cheloveka, Akad. Nauk S.S.S.R., Sibirsk Otdel., Tr. Pervoi Konf., Ulan-Ude, 182–93; (1963). *C.A.* **59**, 8080.
224. Lambin, A. Z. (1959). *Trudÿ Omsk. Sel'.-khoz. Inst.* **37**, 31–9; (1961). *C.A.* **55**, 7737.
225. Pirozhnikov, K. D. (1962). *Zhivotnovodstvo* **24**, 43–4; (1964). *C.A.* **61**, 9999.
226. Okhrimenko, M. F. (1963). Mikroelementy v Zhizni Rast., Zhivotn. i Cheloveka, Akad. Nauk Ukr. S.S.R., Inst. Fiziol. Rast., Tr. Koordinats. Soveshch., 206–9; (1966). *C.A.* **64**, 8877.
227. Okhrimenko, M. F. (1965). Primenenie Mikroelementov v Sel'. Khoz., Akad. Nauk Ukr. S.S.R., 63–8; (1966). *C.A.* **64**, 1306.
228. Akent'eva, L. I. (1966). Mikroelem. Sel'. Khoz. Med. Akad. Nauk Ukr. S.S.R., Respub. Mezhvedom. Sb., 120–5; (1967). *C.A.* **66**, 28070.
229. Bereznitskaya, N. I. and Levenets, P. P. (1966). *Trudÿ Khar'kov. Sel'.-khoz. Inst.* **49**, 186–90; (1968). *C.A.* **68**, 68063.
230. Khristozov, A. and Kanev, S. (1969). *Rastenievud. Nauki* **6**, 87–93; (1970). *C.A.* **72**, 131593.
231. Shevchenko, V. P. (1968). *Nauk. Pr., Ukr. Akad. Sil's'kogospod Nauk* No. 8, 76–82; (1970). *C.A.* **73**, 44375.
232. Barkan, Ya. G. (1964). Mikroelem. Biosfere Ikh Primen. Sel'. Khoz. Med. Sib. Dalnego Vostoka, Dokl Sib. Konf. 2nd, 419–27; (1969). *C.A.* **70**, 86671.
233. Kuznetsov, N. I. (1966). Mikroelem. Zhivotnovod. Rastenievod. Akad. Nauk Kirg. S.S.R., 141–50; (1967). *C.A.* **66**, 64689.
234. Ashkhbabyan, S. A. (1968). *Izv. Sel'.-khoz. Nauk., Min. Sel'. Khoz. Arm. S.S.R.* **11**, 29–33; (1968). *C.A.* **64**, 85781.
235. Noskova, E. I. and Anisimov, A. A. (1970). *Uchen. Zap. Gor'kov. Gos. Univ.* No. 98, 8–13; (1973). *C.A.* **79**, 135831.
236. Astvatsatryan, B. N. and Ashkhbabyan, S. A. (1971). *Khim. Sel'. Khoz.* **9**, 737–9; (1972). *C.A.* **76**, 24016.
237. Rudakova, E. V. (1960). *Nauk Pratsi Ukr. Nauk.-Doslidnii Inst. Fiziol. Roslin, Ukr. Akad. Sil's'kogospod. Nauk* **21**, 70–9; (1962). *C.A.* **56**, 6398.
238. Gyul'akhmedov, A. N. (1961). *Izv. Akad. Nauk Azerb. S.S.R., Ser. Biol. Med. Nauk* No. 4, 65–72; (1962). *C.A.* **57**, 2608.
239. Yakubov, A. M. and Rakhmanov, R. R. (1961). *Uzbek. Biol. Zh.* No. 4, 33–7; (1962). *C.A.* **56**, 10609.
240. Mamedov, Z. I. (1960). *Fiziol. Rast., Akad. Nauk S.S.S.R.* **7**, 724–6; (1961). *C.A.* **55**, 13742.
241. Mamedov, Z. I. (1962). Mikroelementy v Sel'. Khoz. i Med., Ukr. Nauch.-Issled. Inst. Fiziol. Rast. Akad. Nauk Alkr. S.S.R., Materialy 4-go (Chetvertogo) Vses. Soveshch. Kiev, 176–80; (1965). *C.A.* **63**, 6275.
242. Yakubov, A. M. and Uzenbaeva, M. E. (1963). *Khlopkovodstvo* **6**, 46–8; (1963). *C.A.* **59**, 13306.

243. Yakubov, A. M. and Kadyrov, Sh. K. (1963). *Khlopkovodstvo* **13**, 48–50; (1964). *C.A.* **60**, 1065.
244. Solov'ev, V. P., Yaminov, T. I., and Salikhodzhaev, N. A. (1974). *Dokl. Akad. Nauk Uzbek. S.S.R.* **31**, 62–3; (1976). *C.A.* **84**, 73013.
245. Rafikova, G. A. and Shadmanova, S. K. (1973). *Trudy Vses. Nauch.-Issled. Inst. Khlopkovod.* **24**, 205–11; (1974). *C.A.* **81**, 168475.
246. Tashpulatova, A. (1974). *Trudy Vses. Nauch.-Issled. Inst. Khlopkovod.* **26**, 69–70; (1975). *C.A.* **82**, 169357.
247. Kevorkov, A. P. (1961). *Trudy Vses. Nauch.-Issled. Inst. Udobr. Agropochvoved.* No. 38, 286–305; (1963). *C.A.* **58**, 5000.
248. Chepikov, M. S. and Shchetinina, L. L. (1970). *Khim. Sel'. Khoz.* **8**, 495–7; (1970). *C.A.* **73**, 97999.
249. Baryshpol, O. Ya. and Volynskaya, U. M. (1965). Primenie Mikroelementov v Sel'. Khoz. Akad. Nauk Ukr. S.S.R., 120–8; (1965). *C.A.* **63**, 15496.
250. Dobrolyubskii, O. K. (1955). Mikroelementy v Sel'. Khoz. i Med. Akad. Nauk Latv. S.S.R. Otdel Biol. Nauk, Trud. Vses. Soveshch., Riga, 389–400; (1959). *C.A.* **53**, 10628.
251. Gartel, W. (1959). *Weinberg Keller* **6**, 203–10; (1960). *C.A.* **54**, 1786.
252. Kolesnik, L. V. (1965). Primenenie Mikroelementov v Sel'. Khoz. Akad. Nauk Ukr. S.S.R., 203–10; (1965). *C.A.* **63**, 15496.
253. Dobrolyubskii, O. K. (1970). *Agrokhimiya* No. 6, 150–2; (1970). *C.A.* **73**, 108804.
254. Lashkevich, G. I. (1958). Osnovn. Rezul'taty Nauch.-Issled. Raboty Nauch Issled. Inst. Melior i Vodn. Khoz. Akad. Sel'.-khoz. Nauk Beloruss. S.S.R., Minsk Sb., 199–211; (1962). *C.A.* **56**, 6398.
255. Hayakawa, T. and Takayama, T. (1962). *Nat. Inst. Anim. Hlth Quart.* 216–21; (1963). *C.A.* **59**, 8085.
256. Smeltzer, C. G., Langille, W. M., and MacLean, K. S. (1962). *Can. J. Plant Sci.* **42**, 46–52.
257. Powrie, J. K. (1960). *Aust. J. Sci.* **23**, 198–9.
258. Baryshpol, O. Ya. (1960). *Nauk. Pratsi Akad. Sil's'kogospod. Nauk* **17**, No. 12, 84–7; (1962). *C.A.* **57**, 8925.
259. Delwiche, C. C., Johnson, C. M., and Reisenauer, H. M. (1961). *Pl. Physiol.* **36**, 73–8.
260. Danilova, T. A. and Davydova, E. N. (1961). *Dokl. Akad. Nauk S.S.S.R.* **137**, 1470–3; (1961). *C.A.* **55**, 16880.
261. Andreeva, N. M. (1962). Mikroelementy v Sel'. Khoz. i Med. Ukr. Nauch.-Issled. Inst. Fiziol. Rast. Akad. Nauk Ukr. S.S.R., Materialy 4-go (Chetvertogo) Vses. Soveshch., Kiev, 250–4; (1965). *C.A.* **63**, 9008.
262. Petersburgskii, A. V. and Yang, H. (1963). *Dokl. Mosk. Sel'.-khoz. Akad.* No. 84, 213–18; (1964). *C.A.* **60**, 8351.
263. Ozanne, P. G., Greenwood, E. A. N., and Shaw, T. C. (1963). *Aust. J. Agric. Res.* **14**, 39–50.
264. Hallsworth, E. G. and Wilson, S. B. (1962). *Proc. Univ. Nottingham Easter School Agric. Sci.* **9**, 44–6.
265. Tamboli, P. M. and Dube, J. N. (1964). *J. Indian Soc. Soil Sci.* **12**, 371–4.
266. Loercher, L. and Liverman, J. L. (1964). *Pl. Physiol.* **39**, 720–5.
267. Klimov, M. G. and Klimova, L. I. (1966). Mikroelem. Sel'. Khoz. Med. Akad. Nauk Ukr. S.S.R., Respub. Mezhvedom. Sb., 102–5; (1967). *C.A.* **66**, 10269.
268. Okuntsov, M. M. and Kudinova, L. I. (1968). *Mikroelem. Sib.* No. 6, 75–7; (1969). *C.A.* **71**, 29681.

269. Tishchenko, I. V., Altunina, V. A., and Zheleznova, T. A. (1969). *Khim. Sel'. Khoz.* 7, 187–9; (1969). *C.A.* 71, 29674.
270. Sapatyi, S. E. and Shkvaruk, N. M. (1969). *Dokl. Vses. Akad. Sel'.-khoz. Nauk* No. 1, 17–19; (1969). *C.A.* 70, 76853.
271. Ozolina, G. and Zarina, V. (1975). Mikroelem. Komplekse Miner. Pitan. Rast., 117–24; (1976). *C.A.* 84, 29724.
272. Morozov, A. S. (1973). *Trudỹ Vses. Sel'.-khoz. Inst. Zaochn. Obraz.* 68, 37–45; (1976). *C.A.* 85, 4381.
273. Morozov, A. S. and Efimova, T. A. (1966). *Trudỹ Univ. Druzhby Nar.* 14, *Sel'.-khoz. Nauki* No. 1, 43–6; (1967). *C.A.* 67, 21152.
274. Gorodnii, N. M. and Kotik, O. A. (1969). *Visn. Sil's'kogospod. Nauki* 12, 62–4; (1970). *C.A.* 72, 2710.
275. Peive, J., Yagodin, B. A., and Zhiznevskaya, G. Ya. (1966). *Agrokhimiya* No. 9, 53–5; (1966). *C.A.* 65, 17646.
276. Skirpstiene, A. (1966). *Trudỹ Vses. Nauch.-Issled. Inst. Udobr. Agropochvoved.* No. 44, 198–204; (1968). *C.A.* 69, 76021.
277. Kamynina, L. M. and Petersburgskii, A. V. (1967). *Dokl. TSKHA* No. 124, 145–50; (1968). *C.A.* 68, 77284.
278. Petersburgskii, A. V., Kamynina, L. M., Antonova, Z. P., and Nicolov, B. A. (1969). *Dokl. TSKHA* No. 149, 21–6; (1970). *C.A.* 72, 78042.
279. Godnev, T. N. and Leshina, A. V. (1967). *Dokl. Akad. Nauk Beloruss. S.S.R.* 11, 359–61; (1967). *C.A.* 67, 63363.
280. Chuikov, A. G. and Petrakova, L. V. (1967). *Nauch Dokl. Vỹssh. Shk., Biol. Nauki* No. 2, 106–9; (1967). *C.A.* 66, 104385.
281. Shirobokova, E. S. (1972). *Nekot. Vopr. Biol. Fiziol. Rast.* 153–7; (1975). *C.A.* 82, 72076.
282. Klimov, M. G. (1967). *Visn. Sil's'kogospod. Nauki* 10, 68–71; (1967). *C.A.* 67, 90137.
283. Aleksandrova, I. N. (1973). *Trudỹ Gor'kov. Sel'.-khoz. Inst.* 55, 181–6; (1975). *C.A.* 82, 42271.
284. Tishchenko, I. V. (1975). *Agrokhimiya* No. 5, 101–6; (1975). *C.A.* 83, 57376.
285. Chomchalow, S. (1975). *Thai J. Agric. Sci.* 8, 1–5; (1975). *C.A.* 83, 112845.
286. Kevorkov, A. P. (1955). Mikroelementy v Sel'. Khoz. i Med., Akad. Nauk Latv. S.S.R., Otdel. Biol. Nauk, Tr. Vses. Soveshch. Riga, 221–5; (1959). *C.A.* 53, 11539.
287. Ahmed, S. and Evans, H. J. (1969). *Soil Sci.* 90, 205–10.
288. Ahmed, S. and Evans, H. J. (1961). *J. Nat. Acad. Sci. U.S.* 47, 24–36.
289. Ahmed, S. and Evans, H. J. (1962). Radioisotopes Soil–Plant Nutr. Studies, Proc. Symp., Bombay, 259–77; (1963). *C.A.* 59, 1046.
290. Hallsworth, E. G., Wilson, S. B., and Adams, W. A. (1965). *Nature, Lond.* 205, 307–8.
291. Kedrov-Zikhman, O. K. and Rozenberg, R. E. (1955). Trudỹ. Konf. Melior. Osvoen. Bolot, Zaboloch. Pochv. Minsk. Akad. Nauk Beloruss. S.S.R., 374–82; (1959). *C.A.* 53, 12551.
292. Deeva, V. P. (1962). Mikroelementy v Sel'. Khoz. i Med., Ukr. Nauch.-Issled. Inst. Fiziol. Rast., Akad. Nauk Ukr. S.S.R., Materialy 4-go (Chetvertogo) Vses, Soveshch. Kiev, 237–40; (1965). *C.A.* 63, 9007.
293. Danilova, T. A. and Demkina, E. N. (1964). *Agrokhimiya* No. 6, 113–16; (1965). *C.A.* 62, 7060.
294. Al'shevskii, N. G. (1966). *Agrokhimiya* 12, 80–7; (1967). *C.A.* 66, 36897.

295. Al'shevskii, N. G. (1972). *Khim. Sel'. Khoz.* **10**, 344–6; (1972). *C.A.* **77**, 47321.
296. Monakhova, G. P. (1976). *Vest. Sel'.-khoz. Nauki Kaz.* **19**, 21–4; (1976). *C.A.* **85**, 19803.
297. Kuznetsov, N. I. (1964). Mikroelementy v Zhivotnovod. i Rastenievod., Akad. Nauk Kirg. S.S.R., 3–19; (1965). *C.A.* **63**, 1183.
298. Baginskas, B. (1965). Mikroelementy v Sel'. Khoz. Akad. Nauk Uzbek. S.S.R., Otdel. Khim.-Tekhnol. i Biol. Nauk, 187–90; (1966). *C.A.* **64**, 13343.
299. Shmilliar, M. (1968). *Mezögazd. Novenynemesitesi Novenytermesztesi Kut. Intez. Sopronhorpacs, Közl.* **4**, 75–95; (1970). *C.A.* **73**, 3115.
300. Glyadkovskaya, E. I. and Vol'f, Sh. M. (1965). *Vest. Sel'.-khoz. Nauki, Min. Sel'. Khoz. S.S.R* **10**, 92–3; (1966). *C.A.* **65**, 1338.
301. Godnev, T. N. and Leshina, A. V. (1959). Voprosy Fisiol. Rast. i Mikrobiol. Beloruss. Gosudarst. Univ. im V.I. Lenina, 13–17; (1961). *C.A.* **55**, 13744.
302. Efimov, M. V. and Shuin, K. A. (1960). *Trudÿ Buryatsk. Zoovet. Inst.* No. 14, 77–84; (1962). *C.A.* **56**, 10614.
303. Toth, A. and Szabo, V. (1960). *Kisérl. Közl.* **A52**, 47–54; (1963). *C.A.* **59**, 14526.
304. Simakov, N. S. and Gladchuk, V. Ya. (1964). Mikroelm. Biosfere Ikh Primen. Sel'. Khoz. Med. Sib. Dal'nego Vostoka, Dokl. Sib. Konf. 2nd, 408–14; (1969). *C.A.* **70**, 76868.
305. Kulikov, N. V. and Timofeeva, N. A. (1965). *Pochvovedenie* No. 4, 70–4; (1965). *C.A.* **63**, 2348.
306. Abolina, G. I., Ataullaev, N. A., and Azimov, F. (1965). Mikroelementy v Sel'. Khoz. Akad. Nauk Uzbek. S.S.R., Otdel. Khim.-Tekhnol. i Biol. Nauk, 262–73; (1966). *C.A.* **64**, 13343.
307. Shvaruk, N. M. and Sapatyi, S. E. (1964). Mikroelementy v Zhizni Rast., Zhivotn. i Cheloveka Sb., 235–47; (1965). *C.A.* **63**, 13992.
308. Gorid'ko, I. V. (1964). *Fiziol. Rast.* **11**, 922–5; (1965). *C.A.* **62**, 4565.
309. Dubikovskii, G. P., Kudlo, K. K., Talashkevich, V. Ya., and Chertko, N. K. (1970). *Vest. Akad. Navuk Belaruss. S.S.R., Ser. Sel'skagaspad Navuk* No. 1, 47–51; (1970). *C.A.* **73**, 55171.
310. Vecher, A. S. and Satsukevich, V. B. (1972). *Dokl. Akad. Nauk Beloruss. S.S.R.* **16**, 366–9; (1972). *C.A.* **77**, 60670.
311. Shvaruk, N. M. and Nikolaichuk, N. T. (1973). *Nauk. Pratsi. Ukr. Sil's'hogospod. Akad.* **62**, 100–2; (1975). *C.A.* **82**, 30249.
312. Sviridov, A. S. (1974). *Zap. Voronezh. S-kh. Inst.* **61**, 142–6; (1976). *C.A.* **85**, 4359.
313. Yagodin, B. A. (1961). Rol Mikroelementov v Sel'. Khoz. Tr. 2-go (Vtorogo) Mezhvuz. Soveshch. po Mikroelementam, 163–73; (1962). *C.A.* **57**, 10243.
314. Yagodin, B. A. (1963). Rol. Mikroelem. Protsesse Rosta Razv. Rast., Dokl. Nauch. Konf. Vilnyus, 119–22; (1967). *C.A.* **66**, 28092.
315. Isergina, M. M. (1964). *Uchen. Zap. Petrozavodskogo Gos. Univ.* **12**, 15–18; (1965). *C.A.* **63**, 17081.
316. Tsyplenkov, A. E. (1967). Sb. Mater. Vses. Soveshch. Virus. Bolez. Ovoshch. Kul't., 36–40; (1970). *C.A.* **72**, 89348.
317. Dobrolyubskii, O. K., Gaeva, T. A., and Borodin, A. F. (1969). Mater. Nauch. Konf. Fak. Vinograd. Plodovoshchevodstva, Odess. Sel'.-khoz. Inst., 99–109; (1973). *C.A.* **78**, 96541.
318. Ragimov, U. A. (1969). *Khim. Sel'. Khoz.* **7**, 831–2; (1970). *C.A.* **72**, 110358.
319. Pavlenko, I. F. (1969). *Trudÿ Khar'kov. Sel'.-khoz. Inst.* **73**, 27–32; (1970). *C.A.* **73**, 119726.

320. Boryaev, I. M. (1972). *Trudȳ Ul'yanovsk. Sel'.-khoz. Inst.* **17**, 73–94; (1973). *C.A.* **79**, 41368.
321. Tinyakova, N. M. (1972). *Nekot. Vopr. Biol. Fiziol. Rast.*, 129–33; (1975). *C.A.* **82**, 72074.
322. Gorid'ko, I. V. (1972). *Nekot. Vopr. Biol. Fiziol. Rast.*, 41–81; (1975). *C.A.* **82**, 72070.
323. Vaganov, A. P., Knys, A. N., and Sorokina, A. P. (1972). *Trudȳ Khar'k. Sel.-khoz. Inst.* **176**, 7–12; (1976). *C.A.* **84**, 120327.
324. Muzaleva, L. D. (1963). *Uchen. Zap. Petrozavodsk. Univ.* **11**, 10–15; (1965). *C.A.* **62**, 16911.
325. Terent'eva, M. V. (1965). *Botan. Issled. Beloruss. Otdel. Vses. Botan. Obshchestva* No. 7, 36–42; (1966). *C.A.* **64**, 18350.
326. Tsarentseva, M. V. (1959). *Vest. Akad. Navuk Belaruss. S.S.R., Ser. Biyal. Navuk* No. 4, 61–5; (1964). *C.A.* **60**, 7396.
327. Terent'eva, M. V. (1963). *Vest. Akad. Navuk Belaruss. S.S.R., Ser. Biyal. Navuk* No. 2, 56–8; (1965). *C.A.* **62**, 9734.
328. Tsyarents'eva, M. U., Peker, M. Z., and Darozhkina, L. M. (1966). *Vest. Akad. Navuk Belaruss. S.S.R., Ser. Sel'skagaspad. Navuk* No. 4, 21–3; (1967). *C.A.* **66**, 94360.
329. Shuin, K. and Efimov, M. (1960). *Sel'. Khoz. Sibiri* No. 12, 33–5; (1962). *C.A.* **57**, 10243.
330. Yamada, Y. (1958). *Botan. Mag., Tokyo* **71**, 319–25; (1959). *C.A.* **53**, 12416.
331. Russell, S. A., Evans, H. J., and Mayeux, P. (1967). Biol. Alder, Proc. Symp. Northwest Sci. Ass. 40th Ann. Meeting, 259–72.
332. Peskova, R. E. (1968). *Latv. PSR Zināt. Akad. Vest.* No. 7, 111–19; (1968). *C.A.* **69**, 85785.
333. Tammaru, I. (1971). *Tartu Riikliku Ülik. Toim.* No. 270, 19–27; (1972). *C.A.* **76**, 33171.
334. Shabaev, A. I. and Sidorova, V. M. (1969). *Izv. Vȳssh. Ucheb. Zaved., Les. Zh.* **12**, 134–6; (1970). *C.A.* **72**, 54238.
335. Reddy, D. T. and Raj, A. S. (1975). *Pl. Soil* **42**, 145–52.
336. Dobrovol'skii, I. A. and Strikha, E. A. (1971). *Ukr. Botan. Zh.* **28**, 698–702; (1972). *C.A.* **76**, 112003.
337. Nikitin, A. A., Moreva, T. A., and Martinson, T. I. (1964). *Botan. Zh.* **49**, 1294–8; (1965). *C.A.* **62**, 3365.
338. Nesterovich, N. D. and Rakhteenko, L. I. (1966). *Botan. Issled.* No. 8, 106–15; (1967). *C.A.* **66**, 114965.
339. Khashes, Ts. M. and Dolobovskaya, A. S. (1969). *Mikroelem. Sel'. Khoz. Med.* No. 5, 67–72; (1970). *C.A.* **73**, 34303.
340. Vasyaev, G. V. (1969). *Zap. Leningrad. Sel'.-khoz. Inst.* **128**, 28–35; (1970). *C.A.* **73**, 87076.
341. Kasiev, M. Z. and Krasota, V. V. (1970). *Khlopkovodstvo* **20**, 26–7; (1970). *C.A.* **73**, 87048.
342. Lipskaya, G. A., Sergeichik, S. A., Sergeichik, A. A., and Matveentseva, V. S. (1973). *Agrokhimiya* No. 7, 91–7; (1973). *C.A.* **79**, 114492.
343. Lipskaya, G. A. (1974). *Vest. Akad. Navik Beloruss. S.S.R., Ser. Biyal. Navuk* No. 4, 121–2; (1975). *C.A.* **82**, 3308.
344. Lipskaya, G. A. and Zelenaya, L. A. (1975). *Fiziol. Rast.* **22**, 277–81; (1975). *C.A.* **83**, 8152.

345. Shkol'nik, M. Ya. (1961). *Mikroelementy v S.S.S.R., Byull. Vses. Koordinats. Komis. po Mikroelementam* No. 1, 23–9; (1963). *C.A.* **58**, 8381.
346. Lakiza, E. N. (1962). *Dokl. Soobshch. Uzhgorodsk. Gos. Univ., Ser. Biol.* No. 5, 7–8; (1963). *C.A.* **59**, 13304.
347. Dobrolyubskii, O. K., Ryzha, V. K., Fedorenko, I. F., and Pavlenko, M. M. (1962). *Mikroelementy v Sel'. Khoz. i Med., Ukr. Nauch.-Issled. Inst. Fiziol. Rast. Akad. Nauk Ukr. S.S.R., Materialy 4-go (Chetvertogo) Vses. Soveshch.* Kiev, 183–7; (1965). *C.A.* **63**, 7614.
348. Danilova, T. A. (1961). *Trudÿ Vses. Nauch.-Issled. Inst. Udobr. Agropochvoved.* No. 38, 305–12; (1963). *C.A.* **58**, 8381.
349. Shirobokova, E. S. (1972). *Nekot. Vopr. Biol. Fiziol. Rast.*, 158–60; (1974). *C.A.* **82**, 72077.
350. Lipskaya, G. A. (1961). *Vopr. Fiziol. Rast Mikrobiol., Beloruss. Gos. Univ.* No. 2, 20–7; (1963). *C.A.* **58**, 5001.
351. Baginskas, B. (1963). *Rol Mikroelem. Protsesse Rosta Razv. Rast., Dokl. Nauch Konf.* Vilnyus, 15–18; (1967). *C.A.* **66**, 28081.
352. Danilova, T. A. and Popova, M. V. (1964). *Agrokhimiya* No. 2, 106–12; (1964). *C.A.* **61**, 12580.
353. Savchenko, M. P. and Vol'f, Sh. M. (1966). *Agrokhimiya* No. 5, 96–8; (1966). *C.A.* **65**, 6241.
354. Lipskaya, G. A. and Zhaunyarovich, N. I. (1966). *Vest. Akad. Navuk Belaruss. S.S.R., Ser. Biyal. Navuk* No. 2, 31–4; (1966). *C.A.* **65**, 19261.
355. Lipskaya, G. A., Cherkasskaya, S. K., and Matveentseva, V. S. (1972). *Fiziol. Rast.* **19**, 332–5; (1972). *C.A.* **77**, 33498.
356. Lipskaya, G. A. (1974). *Khlorofill*, 380–7; (1975). *C.A.* **82**, 30276.
357. Godnev, T. N. and Leshina, A. V. (1961). *Vopr. Fiziol. Rast. Mikrobiol. Beloruss. Gos. Univ.* No. 2, 13–19; (1963). *C.A.* **58**, 5001.
358. Buzover, F. Ya. (1961). *Mikroelementy i Estestv. Radioaktivn. Pochv. Rostovsk. Gos. Univ. Materialy 3-go (Tret'ego) Mezhvuz. Soveshch.*, 148–9; (1964). *C.A.* **60**, 2288.
359. Yagodin, B. A. (1963). *Nauch. Dokl. Vӳssh. Shk., Biol. Nauki* No. 4, 146–51; (1964). *C.A.* **60**, 7401.
360. Baganov, A. P. (1969). *Trudÿ Khar'kov. Sel'.-khoz. Inst.* **78**, 20–2; (1970). *C.A.* **72**, 99645.
361. Gorid'ko, I. V. (1969). *Fiziol. Rast.* **16**, 405–7; (1969). *C.A.* **71**, 48800.
362. Lipskaya, G. A. and Fartotskaya, I. K. (1970). *Vest. Akad. Navuk Belaruss. S.S.R., Ser. Biyal. Navuk* No. 5, 57–61; (1971). *C.A.* **74**, 31226.
363. Golovina, L. P. and Pavlenko, I. F. (1973). *Agrokhim. Gruntozn.* **22**, 98–103; (1975). *C.A.* **83**, 57334.
364. Duta, A. and Szekely, I. (1973). *An. Univ. Craiova, Ser.3*,**5**, 275–9; (1976). *C.A.* **84**, 42461.
365. Guseinov, B. Z. and Guseinov, S. G. (1961). *Trudÿ Tashkentsk. Konf. po Mirnomu Ispol'z. At. Energii, Akad. Nauk Uzbek S.S.R.* **3**, 222–5; (1962). *C.A.* **57**, 14212.
366. Guseinov, S. G. (1960). *Izv. Akad. Nauk Azerb. S.S.R., Ser. Biol. Med. Nauk* No. 4, 3–9; (1962). *C.A.* **56**, 15846.
367. Bunyatov, I. M. (1961). *Uchen. Zap. Azerb. Gos. Univ. Ser. Biol. Nauk* **6**, 31–3; (1963). *C.A.* **59**, 7860.
368. Bunyatov, I. M. (1960). *Uchen. Zap. Azerb. Gos. Univ. im S.M. Kirova, Ser. Biol. Nauk* No. 3, 23–7; (1962). *C.A.* **56**, 9143.

369. Efimov, M. V. (1962). *Mikroelementy v Vost. Sibiri i na Dal'n. Vost., Inform. Byull. Koordinats. Komis. po Mikroelementam dlya Sibiri i Dal'n. Vostoka* No. 1, 76–8; (1963). *C.A.* **59**, 15911.
370. Lipskaya, G. A., Matveentseva, V. S., and Sergeichik, S. A. (1972). *Dokl. Akad. Nauk Beloruss. S.S.R.* **16**, 70–2; (1972). *C.A.* **76**, 139432.
371. Kostir, J. and Jiracek, V. (1959). *Naturewissenschaften* **46**, 270; (1959). *C.A.* **53**, 19053.
372. Gavrilova, P. N. (1965). *Trudy Saratov. Zootekh.-Vet. Inst.* **13**, 56–66; (1967). *C.A.* **66**, 94359.
373. Enileev, Kh. Kh. and Andryushchenko, V. K. (1963). *Uzbek. Biol. Zh.* **7**, 23–7; (1964). *C.A.* **60**, 11329.
374. Mekhti-Zade, R. M. and Lyatifov, D. Kh. (1962). *Dokl. Akad. Nauk Azerb. S.S.R.* **18**, 29–33; (1963). *C.A.* **58**, 11915.
375. Morozov, A. S. (1966). *Trudy Univ. Druzhby Nar.* 14, *Sel'.-khoz. Nauki* No. 1, 40–2; (1967). *C.A.* **67**, 21151.
376. Kamynina, L. M. (1965). *Agrokhimiya* No. 10, 123–7; (1966). *C.A.* **64**, 7315.
377. Shklyaev, Ya. N. (1967). Sb. Aspir. Rab. Kazan. Gos. Univ., Biol., 75–9; (1969). *C.A.* **70**, 76881.
378. Stanchev, L. and Gorbanov, S. (1969). *Pochvozn. Agrokhim.* **4**, 125–32; (1970). *C.A.* **72**, 131585.
379. Mel'nikova, N. I., Petrushin, V. V., Sitnikov, V. V., and Ryzhkov, V. S. (1969). *Trudy Kostromskogo Sel'.-khoz. Inst.* No. 21, 103–7; (1972). *C.A.* **76**, 24023.
380. Guseinov, S. G. (1959). *Izv. Akad. Nauk Azerb. S.S.R., Ser. Biol. Sel'.-khoz. Nauk* No. 5, 81–8; (1961). *C.A.* **55**, 9757.
381. Soboleva, A. V. (1963). *Fiziol. Rast.* **10**, 719–21; (1964). *C.A.* **60**, 11327.
382. Ikeda, M. and Kurozumi, K. (1967). *Hiroshima Daigaku Suichikusan Gakubu Kiyo* **7**, 149–70; (1968). *C.A.* **69**, 18305.
383. Potakhina, L. N. (1965). *Uchen. Zap. Petrozavodsk. Gos. Univ.* **13**, 130–1; (1966). *C.A.* **65**, 19261.
384. Dobrolyubskii, O. K. and Slavvo, A. V. (1960). *Biokhim. Vinodeliya Sb.* **6**, 196–222; (1961). *C.A.* **55**, 7739.
385. Mekhti-Zade, R. M. and Lyatifov, D. Kh. (1963). Mikroelementy v Sel'.-khoz. i Med. Sb., 286–91; (1965). *C.A.* **62**, 828.
386. Mironova, M. P. and Kholoptseva, N. P. (1964). *Uchen. Zap. Petrozavodsk. Gos. Univ.* **12**, 32–5; (1966). *C.A.* **65**, 6247.
387. Ananyan, A. M., Garibyan, G. A., Avakyan, A. G., and Avundzhan, E. S. (1976). *Izv. Sel'.-khoz. Nauk* **19**, 44–52; (1976). *C.A.* **85**, 141913.
388. Pavel, J. and Zakova, J. (1967). *Biol. Plant* **9**, 383–91; (1968). *C.A.* **68**, 2310.
389. Satsukevich, V. B. (1974). *Dokl. Akad. Nauk Beloruss. S.S.R.* **18**, 271–4; (1974). *C.A.* **80**, 144762.
390. Satsukevich, V. B. (1975). *Vest. Akad. Navuk B.S.S.R., Ser. Biyal. Navuk* No. 3, 98–101; (1975). *C.A.* **83**, 77524.
391. Agarwala, S. C. and Kumar, A. (1962). *J. Indian Botan. Soc.* **41**, 77–92.
392. Potakhina, L. N. (1965). *Uchen. Zap. Petrozavodsk. Gos. Univ.* **13**, 40–2; (1966). *C.A.* **65**, 19260.
393. Yagodin, B. A. and Petrovskaya, G. V. (1971). Nov. Izuch. Biol. Fiksatsii Azota, 42–7; (1972). *C.A.* **76**, 112035.
394. Sergeeva, N. V., Brun, G. A., and Drozdova, I. V. (1971). *Izv. Akad. Nauk Mold. S.S.R., Ser. Biol. Khim. Nauk* No. 5, 54–7; (1972). *C.A.* **77**, 4235.

395. Anisimov, A. A. and Ganicheva, O. P. (1973). *Uchen. Zap. Gor'kov. Gos. Univ.* **178**, 49–51; (1975). *C.A.* **83**, 8171.
396. Anisimov, A. A. and Noskova, E. Z. (1973). *Uchen. Zap Gor'kov. Gos. Univ.* **178**, 52–7; (1975). *C.A.* **83**, 8172.
397. Lapa, V. G., Romanchuk, P. S., Skuratovskaya, M. Z., and Levina, N. I. (1975). Pochv. Usloviya i Effektivnost Udobr. No. 1, 131–41; (1965). *C.A.* **63**, 1185.
398. Tinyakova, N. M. (1972). *Nekot. Vopr. Biol. Fiziol. Rast.*, 134–7; (1975). *C.A.* **82**, 72075.
399. Satsukevich, V. B. (1972). *Vest. Akad. Navuk Beloruss. S.S.R.*, Ser. Biyal Navuk No. 2, 42–6; (1972). *C.A.* **77**, 33408.
400. Usik, G. E. and Beskrovnaya, V. N. (1969). *Mikroelem. Sel'. Khoz. Med.* No. 5, 106–12; (1970). *C.A.* **73**, 44369.
401. Peive, J. (1964). *Agrokhimiya* No. 7, 3–18; (1965'. *C.A.* **62**, 6809.
402. Fujii, R. (1962). *Botan. Mag., Tokyo* **75**, 176–84; (1963). *C.A.* **59**, 10465.
403. Steklova, M. M. (1963). *Trudÿ Botan. Inst., Akad. Nauk S.S.S.R., Eks. Botan.*, Ser. 4, No. 16, 3–26; (1963). *C.A.* **59**, 9079.
404. Sironval, C. (1962). *C. R. Congr. Soc. Savantes Paris Sect. Sci.* **87**, 1135–40; (1964). *C.A.* **61**, 959.
405. Petrov, A. I. (1961). *Trudÿ Karel'sk. Filiala Akad. Nauk S.S.S.R.* No. 29, 50–8; (1962). *C.A.* **57**, 12922.
406. Mekhti-Zade, R. M. and Lyatifov, D. Kh. (1961). *Uchen. Zap. Azerb. Gos. Univ.*, Ser. Biol. Nauk No. 3, 45–51; (1963). *C.A.* **58**, 7323.
407. Gorid'ko, I. V. (1967). *Nauch Dokl. Vÿssh. Skh., Biol. Nauki* No. 3, 84–7; (1967). *C.A.* **67**, 10807.
408. Semina, R. M. (1970). *Biol. Nauki* No. 6, 69–72; (1971). *C.A.* **74**, 52540.
409. Kedrov-Zikhman, O. K. (1955). Primenenie Izotop. Tekh., Biol. Sel'. Khoz., 349–65; (1959). *C.A.* **53**, 12409.
410. Kedrov-Zikhman, O. K., Rozenberg, R. E., and Protashchik, L. N. (1955). Mikroelementy v Sel'. Khoz. i Med., Akad. Nauk Latv. S.S.R., Otdel. Biol. Nauk., Tr. Vses. Soveshch. Riga, 51–65; (1959). *C.A.* **53**, 10621.
411. Kedrov-Zikhman, O. K., Kozhevnikova, A. N., and Protashchik, L. N. (1955). Primenenie Izotopov pri Agrokhim. i Pochvennykh Issledovan., Akad. Nauk S.S.S.R. Pochvennyi Inst. im V.V. Dokuchaeva, Sbornik Statei, 242–70; (1959). *C.A.* **53**, 13477.
412. Abutalybov, M. G. (1959). *Uchen. Zap. Azerb. Gos. Univ., Biol. Ser.* No. 3, 33–42; (1963). *C.A.* **58**, 2808.
413. Abutalybov, M. G. and Aliev, D. A. (1960). *Uchen. Zap. Azerb. Gos. Univ.*, Ser. Biol. Nauk No. 5, 27–34; (1962). *C.A.* **57**, 12922.
414. Abutalybov, M. G. and Aliev, D. A. (1961). *Izv. Akad. Nauk Azerb. S.S.R.*, Ser. Biol. Med. Nauk No. 5, 31–40; (1962). *C.A.* **56**, 2723.
415. Tiffin, L. O. (1967). *Pl. Physiol.* **42**, 1427–32.
416. Petersburgskii, A. V. and Sadovskaya, E. N. (1972). *Dokl. TSKHA* No. 188, 119–23; (1974). *C.A.* **80**, 36241.
417. Bozhenko, V. P. and Shkol'nik, M. Ya. (1963). Vodn. Rezhim Rast. v Svyazi s Obmenon Veshchestv i Produktivnost'yu, Akad. Nauk S.S.S.R., Inst. Fiziol. Rast., 275–83; (1964). *C.A.* **60**, 12619.
418. Bozhenko, V. P., Nazarenko, A. M., and Momot, T. S. (1963). Mikroelementy v Sel'. Khoz. i Med. Sb., 168–72; (1965). *C.A.* **62**, 9732.
419. Shkol'nik, M. Ya., Bozhenko, V. P., and Maevskaya, A. N. (1960). Fiziol. Ustoichivosti Rast. Sb., 522–7; (1962). *C.A.* **57**, 10239.

420. Bozhenko, V. P., Shkol'nik, M. Ya., and Momet, T. S. (1963). *Dokl. Akad. Nauk S.S.S.R.* **153**, 1447–9; (1964). *C.A.* **60**, 12617.
421. Bozhenko, V. P. (1968). *Fiziol. Rast.* **15**, 116–22; (1968). *C.A.* **68**, 94925.
422. Mursaliev, A. (1973). *Mikroelem. Zhivotnovod. Rastenievod.* **12**, 67–76; (1975). *C.A.* **83**, 42099.
423. Eyubov, I. Z. (1965). *Trudy Azerb. Nauch.-Issled. Vet. Inst.* **19**, 231–6; (1967). *C.A.* **67**, 21077.
424. Lyon, G. L., Brooks, R. R., Peterson, P. J., and Butler, G. W. (1968). *Pl. Soil* **29**, 225–40.
425. Clader, A. B. and Voss, R. C. (1957). *Bull. Consult. Comm. Dev. Spectrogr. Wk* No. 1; (1960). *C.A.* **54**, 25438.
426. Zakirov, K. Z., Rish, M. A., and Ezdakov, V. I. (1959). *Uzbek. Biol. Zh., Akad. Nauk Uzbek. S.S.R.* No. 1, 15–20; (1959). *C.A.* **53**, 13276.
427. Zyka, V. (1974). *Sb. Geol. Ved. Technol., Geochem.* **12**, 157–60; (1975). *C.A.* **82**, 123913.
428. Luzzati, A. (1963). *Annali Sper. Agra.* **17**, 183–220; (1966). *C.A.* **64**, 8880.
429. Anderson, A. J., Meyer, D. R., and Mayer, F. K. (1973). *Aust. J. Agric. Res.* **24**, 557–71.
430. Minami, K., Yasuda, T., and Araki, K. (1973). *Tokai Kinki Nogyo Shikenjo Kenkyu Hokoku* **25**, 48–56; (1974). *C.A.* **80**, 132095.
431. Chino, M. and Kitagishi, K. (1966). *Nippon Dojo-Hiryogaku Zasshi* **37**, 342–7; (1966). *C.A.* **65**, 16018.
432. Yagodin, B. A. and Zhiznevskaya, G. Ya. (1969). *Fiziol. Rast.* **16**, 505–11; (1969). *C.A.* **71**, 69767.
433. Venkatasubramanyam, V., Adiga, P. R., Sastry, K. S., and Sarma, P. S. (1962). *J. Sci. Ind. Res.* **21C**, 167–70.
434. Small, H. G., Sherbek, T. G., Bauer, M. E., and Ohlrogge, A. J. (1967). *Agron. J.* **59**, 564–6.
435. Hara, T., Sonoda, Y., and Iwai, I. (1976). *Soil Sci. Pl. Nutr.* **22**, 317–25; (1976). *C.A.* **85**, 158692.
436. Patel, P. M., Wallace, A., and Mueller, R. T. (1976). *J. Am. Soc. Hortic. Sci.* **101**, 553–6.
437. Guseinov, B. Z. and Guseinov, S. G. (1961). *Trudy Tashkent. Konf. Mirnomu Izpol'z. At. Energii, Akad. Nauk Uzbek. S.S.R.* **3**, 262–7; (1962). *C.A.* **57**, 14202.
438. Khakimov, Kh. and Alizhanov, A. (1974). *Trudy Tashkent. Sel'.-khoz. Inst.* **47**, 26–33; (1976). *C.A.* **84**, 3622.
439. Yakubov, A. M. and Usmanov, Kh. U. (1959). Khim. Khlopchatnika, Akad. Nauk Usbek. S.S.R., Inst. Khim. Rastitel. Veschchestv., 48–57; (1961). *C.A.* **55**, 14787.
440. Timoshenko, A. G. (1959). *Trudy Kishinev Sel'.-khoz. Inst.* **20**, 109–27; (1962). *C.A.* **56**, 2725.
441. Barmenkov, Ya. P. and Shtark, P. A. (1961). *Trudy Orenburgsk. Sel'.-khoz. Inst.* **12**, 379–85; (1963). *C.A.* **58**, 9587.
442. Abdullaeva, K. (1959). *Sotsialist. Sel'. Khoz. Aberbaidzhana* No. 2, 20–2; (1960). *C.A.* **54**, 20046.
443. Enari, T. M. and Kauppinen, V. (1961). *Acta Chem. Scand.* **15**, 1513–16.
444. Mashev, N. and Kutacek, M. (1976). *Pochvozn. Agrokhim.* **11**, 55–63; (1976). *C.A.* **85**, 75013.
445. Herich, R. and Bobak, M. (1976). *Experientia* **32**, 570–1; (1976). *C.A.* **85**, 940.
446. Grover, S. and Purves, W. K. (1976). *Pl. Physiol.* **57**, 886–9.

447. Lundblat, K. (1959). Kal. Lantbrukskogskol. och Statens Lantbruksforsok
     Statens Jordbruksforsok Medd. No. 99; (1961). *C.A.* **55**, 4845.
448. Bluzmanas, P. and Stakauskaite, R. (1962). *Liet. TSR Aukstuju Mokyklu
     Mokslu Darb., Biol.* **2**, 5–24; (1963). *C.A.* **59**, 10728.
449. Vostrilova, N. V. and Dulova, V. I. (1958). *Akad. Nauk S.S.S.R.* No. 1, 69–79;
     (1959). *C.A.* **53**, 9384.

8 Cobalt in Animal Nutrition

Cattle and Sheep

The discovery that minute quantities of cobalt are necessary for the maintenance of health in cattle and sheep is an illustration of the time lag which frequently occurs between observation of a natural phenomenon and its ultimate scientific explanation. For a very long time, farmers and ranchers in various parts of the world knew that if cattle and sheep were grazed continuously in certain areas the animals would lose appetite and weight, become anemic and weak, and finally die. These disorders, known by local names such as "bush sickness" in New Zealand, "coast disease" in Australia, "pining" in Britain, and "salt sick" in Florida, were characterized by broadly similar symptoms. The animals could be restored to normal health by being moved to other pastures on another type of soil or overlying a different rock formation.

For many years, agricultural scientists endeavored to find an explanation by seeking toxic elements in the soil and vegetation, parasitic infestation peculiar to the affected region, and deficiencies in the soils and forages of the major elements essential for animal nutrition. When these studies failed to find the cause of the malady, workers in New Zealand and Australia, where the economic importance of this problem was greatest, turned to a study of minor mineral elements, such as iron. Conflicting results were obtained by iron therapy until it was finally discovered, around 1934, that the small quantity of cobalt present in some iron compounds was actually the curative agent. It was soon proved that cattle and sheep could be kept healthy on the formerly "unhealthy" pastures by adding small quantities of cobalt to feed, salt-licks, water, or top dressings for herbage.

The pioneer investigations of Aston,[1] Askew,[2] Rigg,[3] Kidson,[4,5] Dixon,[6] McNaught,[7,8] and others in New Zealand, and of Filmer,[9] Underwood,[10,11] Marston and co-workers[12,13] in Australia, have been a tremendous aid to world agriculture. The significance of cobalt deficiency is well emphasized by Sir Theodore Rigg's statement that in New Zealand alone, as a result of work on cobalt, hundreds of thousands of acres now support cattle and sheep where formerly it was impossible.[3]

The conclusions of those working in the Antipodes on the corrective action of cobalt have been substantiated in the following countries where similar deficiency diseases were found: Brazil, Britain, Canada, Czechoslovakia, Denmark, Eire, Finland, Germany, Holland, Hungary, India, Italy, Japan, Kenya, Poland, South Africa, Sweden, U.S.A., U.S.S.R., and Yugoslavia.

Cobalt deficiency was long considered a disease affecting only cattle and sheep, because horses can apparently graze indefinitely, or certainly for extended periods, on deficient pastures without ill effects. Although ruminants may be the only animals that manifest specific disorders when confined to low-cobalt feeds, a voluminous literature over the past twenty years attests to the fact that the addition of small quantities of cobalt to the rations of many animals and birds has improved growth and development.

In this chapter, the role of cobalt in cattle and sheep will be examined in some detail, followed by a summary of our present knowledge on the effect of this element in other animals, and in birds, fish, and insects. The subject of cobalt in animal nutrition has been reviewed on many occasions.[14-30]

Cobalt content in plants and soils of deficient areas

The results of investigations on the minimum quantity of cobalt required in the pasture herbage or feedstuffs for the maintenance of health in ruminants have varied slightly. Earlier work showed that a cobalt content in the forage below $0.08-0.10$ mg kg^{-1} might produce deficiency diseases in cattle and sheep.[19,31] In recent years, the critical minimum has been frequently placed at 0.07 mg cobalt kg^{-1} of dry feed.[32-37] Two papers suggest $0.04-0.07$ mg cobalt kg^{-1},[32,38] and another found a deficiency at 0.06 ± 0.04, whereas healthy pastures had 0.20 ± 0.07 mg cobalt kg^{-1} of dry feed.[39]

Several publications have favored a slightly higher minimum. In the Pskov area of the U.S.S.R., $0.16-0.27$ mg cobalt kg^{-1} of winter rations is considered a low level,[40] and in the Bashkir region of that country, where the cobalt content in hay ranged from $0.11-0.79$ mg kg^{-1}, a cobalt supplement was required in some areas.[41] In Baden, 0.24 mg cobalt kg^{-1} of dry hay was considered too low for the requirements of milk cows and fattening calves.[42]

To summarize, most studies have indicated that a cobalt deficiency in ruminants may occur when the feedstuffs contain less than $0.07-0.1$ mg cobalt kg^{-1} of dry matter. Many areas of the world have reported such cobalt-deficient herbage or fodder.[19,32-39,43-50] A tabulation of a number of commercial and farm feeds also illustrates the differences which may be expected.[51]

The minimum level of cobalt in the soil required to yield at least about $0.07-0.08$ mg cobalt kg^{-1} dry matter in herbage or fodder crops is, understandably, a debatable figure. Cobalt-deficient soils have been reported to contain, as total cobalt, $0.6–2.3$ mg kg^{-1} of dry soil,[52] 2.5 mg kg^{-1},[37,53] and 5 mg kg^{-1},[35] whereas the total cobalt content of healthy soils has been quoted as 5 mg kg^{-1},[53] and 6.7 mg kg^{-1}.[54] Available cobalt, which is that portion soluble in dilute acid or salt solutions, as listed earlier in Chapter 2, has been reported for deficient soils as 0.17 mg kg^{-1} of soil,[55] and $0.1–0.76$ mg kg^{-1},[50] whereas healthy soils have contained 0.3 mg kg^{-1},[56] and 3 mg kg^{-1}.[57]

In general, a soil content of less than approximately 5 mg total cobalt kg^{-1} of dry soil, or about 0.3 mg available cobalt kg^{-1} of dry soil, may not provide the necessary minimum cobalt content of around $0.07–0.08$ mg kg^{-1} in herbage or fodder for the maintenance of health in cattle and sheep.

An observation on an unexpected occurrence of cobalt deficiency in Eire is worthy of note. Soil cobalt was adequate but plant levels were low. The manganese content of the soil was high, and apparently this element inhibited the plant uptake of cobalt.[58] In Australia, it has been reported that clover-disease symptoms in ewes on pastures over cobalt-deficient soil were exacerbated in the group receiving the cobalt supplement.[59] In Western Australia, cobalt deficiency is worst in the karri-soil areas of the southwest, and the calcareous littoral.[60]

Effect of cobalt supplements on the growth of ruminants

1. *Cattle*

Many publications have recommended varying quantities of cobalt salts as feed supplements to prevent deficiency diseases in areas where the soil and forage are low in cobalt. The growth and development of both calves and cows are enhanced by the addition to their ration of cobalt chloride and other cobalt salts.[61–64] Increased growth has been reported by cobalt chloride supplements of 0.1 mg kg^{-1} of feed,[65] and 0.18 mg kg^{-1} of feed.[66] Cows weighing $450–500$ kg, with average daily milk yields of $10–12$ kg, required $4.3–4.9$ mg of cobalt daily.[67] Other workers have expressed these requirements in a different manner: daily supplement per head of 5 mg cobalt chloride;[68] daily supplement per head of 20 mg cobalt chloride;[69] daily intake of cobalt per head $9–13$ mg;[70] and a daily retention in cows of $2.2–9.7$ mg of cobalt.[71] One paper reported that a cobalt level of 253 mg kg^{-1} in the feed was the optimum for weight gain in young steers, although levels of 152 and 754 mg kg^{-1} also produced good growth.[72] These figures appear extremely high, but not impossible, because it has been shown that growing cattle can consume up to 50 mg cobalt per 45 kg body weight without ill effects.[73] Other workers have indicated that cobalt supplements should be provided if the basic ration for cows does not contain $1.2–1.7$ mg cobalt kg^{-1} of feed,[74] or 2 mg cobalt kg^{-1} of feed for calves,[75] or 0.1 mg cobalt kg^{-1} of feed dry matter for calves,

young cattle, and milk cows.[76] We can conclude this paragraph by quoting
the opinion of Mitchell,[56] an investigator with long experience in the field of
cobalt deficiencies, that a cobalt application of 1–2 kg ha^{-1} may control
cobalt deficiency in grazed pastures. Depending on the availability of soil
cobalt, such applications are presumably effective for 3–5 years.[19]

A publication from Australia on cobalt supplementation showed that no
response was elicited,[77] and a contribution from the U.S.S.R. indicated that
increased uptake of cobalt did not affect either milk production or the quality
of calves.[78] In such instances, the natural fodders obviously provided
sufficient cobalt for the normal growth of cattle.

2. Sheep

About 2·4 kg cobalt sulphate ha^{-1} applied to cobalt-deficient pastures,
increased the cobalt content of herbage from 0·09–0·38 mg kg^{-1}, and pre-
vented cobalt defciency in lambs.[79] The addition to sheep feed of 0·5–1 mg
cobalt chloride kg^{-1} dry matter resulted in increased growth.[80] Supplemental
cobalt is required if the cobalt content of feed is less than 0·08 mg kg^{-1},[81]
0·22–0·25 mg kg^{-1},[82] 0·30 mg kg^{-1},[83] or 0·60 mg kg^{-1}.[84] Addition of cobalt
improved the growth and development of sheep, stimulated lamb growth, and
increased wool yield.[85–87]

Supplements of differing cobalt content have been reported to correct
cobalt deficiencies: daily additions per sheep of 0·15 mg cobalt,[88] 1 mg
cobalt,[89,90] 1 mg cobalt chloride,[91] 1–2 mg cobalt chloride,[92] 2 mg cobalt
chloride,[93,94] 2·5 mg cobalt chloride,[52] 3–6 mg cobalt sulphate,[95,96] 4 mg
cobalt chloride,[97–99] 6 mg cobalt sulphate,[100–101] 10 mg cobalt chloride,[102]
and 5–20 mg cobalt chloride.[103]

Cobalt deficiency was induced in lambs by a diet containing only 0·04 mg
cobalt per day.[104] In another study of hypocobaltosis in sheep, lowering the
cobalt intake from 2 mg to 0·3 mg per day brought about the onset of
deficiency symptoms.[105]

An occasional reference shows that no benefit was gained from addition of
cobalt. When sheep rations in Russia contained 0·128–0·386 mg cobalt kg^{-1},
an addition of 2–10 mg cobalt chloride per day showed no effect;[106] likewise,
in South Africa, the addition of about 1 mg cobalt chloride per day to sheep
rations failed to produce any benefit.[107] In both cases, the cobalt content of
the forage was above the critical minimum required to maintain sheep in good
health.

Cobalt in meat, organs, and blood of ruminants

1. Cattle

One investigator reported that the cobalt content of beef increased with age, with veal containing 0·003 mg cobalt kg^{-1} of meat, and beef 0·011 mg kg^{-1}.[108] In a large number of samples of cured meats, cobalt varied from 0–0·41 mg kg^{-1}, with an average of 0·06 mg.[109] The cobalt content in liver has been reported to be higher than in other organs;[110] liver from normal cattle has been found to contain 0·076 mg cobalt kg^{-1},[111] 0·135 mg kg^{-1},[112] and 0·201 mg kg^{-1}, and liver from animals suffering from cobalt deficiency contained 0·056 mg cobalt kg^{-1}.[113] The results of one examination of cattle liver and pancreas showed that cobalt was generally absent, but, when present, occurred to the extent of 10–30 mg kg^{-1} of the ash.[114] In cows, cobalt is present in the blood at reported values, in mg cobalt kg^{-1} of blood, of 0·025,[115] 0·0315 for fetal blood, and 0·0213 for maternal blood,[116] 0·1 for sterile cows,[117] 0·041–0·084 for dry cows and 0·128–0·233 for pregnant cows.[118]

The cobalt content of the myocardium of cattle has been given at 0·033 mg cobalt kg^{-1}, and that of calves at 0·022 mg.[119] In dairy cattle, a correlation was obtained between cobalt and zinc in the blood.[120]

2. Sheep

The cobalt content of sheep liver has been reported, in mg kg^{-1}, as 0·112,[111] 0·148,[112] 0·04–0·05 for normal animals and 0·001–0·002 for unhealthy sheep,[121] up to 0·43 in liver and up to 1·45 in spleen,[54] and 0·60–1·00 in the livers of wool-bearing sheep.[122]

The blood of sheep was found to contain varying quantities of cobalt, depending on pasture, location, type of animal, and other factors; in mg kg^{-1}, it ranged from 0·016–0·088.[123] The blood of milking sheep contained three times the cobalt of non-milking animals,[124] and the blood of sheep exposed to chronic lead intoxication had a low cobalt content.[125]

Cobalt in ruminant hair and wool

Italian workers have shown that the cobalt content of cattle hair varied from 0·179–0·234 mg kg^{-1}, depending on location and breed.[126] German investigators reported that the black hair of beef cattle contained more cobalt than did the white hair, and that the cobalt content of hair reflected the content of the feed.[127]

It has been frequently demonstrated that cobalt additions to the feed of sheep not only resulted in weight gains, but also in improved wool yields.[80,89,94,95] One study has shown that the addition of cobalt chloride to sheep rations gave an increase of 2–3 times in the solid substance of normal wool; black wool contained more cobalt than did white wool.[128]

Cobalt in milk

The range of cobalt contents in milk is wide, obviously reflecting differences in feed, season, the breed and peculiarities of dairy cows, and other factors. Papers have reported cobalt, in mg l^{-1}, or parts per million, 0·00084,[129] 0·010,[112] 0·016,[111] 0·082–0·208,[130] 0·034–0·041,[131] and 0·072–0·124.[132]

The literature furnishes abundant evidence of the beneficial effect of cobalt supplements on milk production. From 10–40 mg cobalt chloride per head per day increased milk yield.[133–144] One worker, however, while finding that cobalt additions to the feed greatly increased the cobalt content of the milk, was unable to demonstrate any clear effect on milk yield.[145]

It has been observed that the cobalt content of milk drops gradually at the beginning of the lactation period and increases towards the end.[146] A paper showed that the superior cobalt content of fodder in a steppe region over a non-forested area was reflected in both the quantity and quality of milk.[147] Another worker indicated that cows receiving a cobalt supplement used their rations more efficiently, the result being higher milk production.[148]

Two publications have stated that cobalt additions have increased the cobalt content of colostrum and improved its nutritional qualities.[149,150] One worker found that the addition of cobalt to cattle fodder at the rate of 2 mg cobalt 100 kg^{-1} live weight per day, accelerated the ripening of cheese and improved its quality.[151]

The physiological effect of cobalt on milk has been examined by two investigators. The addition of 0·1 mg cobalt kg^{-1} of feed increased the sugar and total nitrogen, but decreased the content of acetoacetic and β-hydroxybutyric acids, ammonia, and urea.[152] A supplement of 20 mg cobalt chloride kg^{-1} of feed raised the levels of both glycoproteins and neuraminic acid.[153]

Effect of cobalt on digestibility in ruminants

The addition of cobalt to feedstuffs of cattle and sheep has improved the digestibility of nutrients. Supplements of 0·2 mg cobalt chloride kg^{-1} of body weight of calves,[154] 40 mg cobalt chloride per day for milk cows,[155] and 5 mg cobalt chloride per head per day for sheep,[156] increased the digestibility and utilization of food.

There is some evidence that the addition of cobalt chloride to the feed of cattle and sheep stimulates rumen metabolism, and activates the microflora for digesting cellulose.[157–159] However, other workers found that cobalt had no appreciable effect on fermentation processes in the rumen and digestion in cattle,[160] and that 5 parts 10^{-6} cobalt inhibited fatty acid production from cellulose, starch, and glucose.[161]

Administration of cobalt in pellets to ruminants

The correction of cobalt deficiency in cattle and sheep has sometimes been effected by the administration of cobalt-containing pellets or granules, or "bullets". These are usually an oxide of cobalt mixed with china clay, and are given directly to the animal as a pill. Good results have been reported by workers in Australia, New Zealand, Kenya, and Russia.[162–168] On the other hand, cobalt supplementation using pellets has failed to produce benefits both in Nebraska[169] and in Queensland.[170]

Cobalt and vitamin B_{12} in ruminants

The relationship between cobalt and vitamin B_{12} was briefly summarized in Chapter 1. The effect of cobalt on the occurrence and function of vitamin B_{12} in ruminants will be examined in more detail in the following paragraphs.

It was found some years ago that cobalt deficiency is associated with ruminant livers in which the vitamin B_{12} concentration is less than 0.1 parts 10^{-6}, and that cobalt additions to the diet substantially increased the vitamin B_{12} content in livers,[171–173] kidneys,[173] and colostrum and milk.[165] Later work has extended our knowledge on the effect of cobalt additions on vitamin B_{12} production, and it is suggested that 0.2 parts 10^{-6} vitamin B_{12} in livers of lambs be the minimum concentration associated with freedom from cobalt deficiency.[174,175]

The following daily additions of cobalt have been reported to raise the level of vitamin B_{12} to satisfactory levels and overcome cobalt deficiencies: 2–4 mg cobalt chloride for sheep,[176] 3–6 mg cobalt sulphate for sheep,[96] 2.6–11.7 mg cobalt for cattle,[177] 5 mg cobalt chloride for sheep,[178] 20 mg cobalt chloride for cows,[179] and 30 mg cobalt chloride for cows.[180] Daily cobalt intakes of 0.047, 0.41, and 0.83 mg Co gave B_{12} productions of 37, 1006, and 1553 μg, respectively.[181]

Investigators have observed that cobalt improved the accumulation of B_{12} in silage corn,[182] and both blood and plasma.[87,183,184] One worker reported that cobalt levels above 5 parts 10^{-6} inhibited vitamin B_{12} synthesis by rumen bacteria.[185] Another paper concluded that the concentration of B_{12} in milk, liver, and gluteal muscles of sheep depended not only on the amount of cobalt in the ration, but also on iodine content and the level of general nutrition.[186] It was found that as many as nine distinct cobamides and cobinamides were detected in sheep rumen contents,[187] and the vitamin B_{12} activity of the latter increased within 2 h of adding cobalt.[188]

One investigator stated that the only demonstrated role of cobalt in living organisms is its participation in the structure of vitamin B_{12}, and that the minimum daily dose of cobalt for ruminants is 0.1 mg.[189] One publication showed that sheep on a low cobalt diet had a low plasma ascorbic acid concentration, indicating a vitamin C-deficient condition; vitamin B_{12} treatment slowly increased the concentration of plasma ascorbic acid.[190]

**Cobalt in the biochemistry and
physiology of ruminants**

Many publications attest to the interest in the influence of cobalt on many aspects of the biochemistry and physiology of cattle and sheep. Cobalt additions have increased the concentration of hemoglobin and erythrocytes in the blood;[65,67,89,92,93,97,99,103,154,155,167,191-197] one paper reported no influence from cobalt.[91]

Cobalt supplements to ruminants have either increased the protein of blood and serum, or had favorable effects on the intensity of nitrogen metabolism.[52,67,85,100,144,154,194,195,197-205] One paper reported that blood levels of nitrogen-containing compounds in cobalt-deficient lambs were higher than normal in the middle stage, but returned to normal at the end of the experiment.[206] Another worker found that nitrogen levels in sheep muscle were about the same during malnutrition and cobalt deficiency.[207] It was reported that the addition of cobalt chloride to cattle in a cobalt-deficient area did not produce any consistent change in blood serum protein.[137]

A few experiments on the carbohydrate metabolism of ruminants have been recorded.[154,208-210] Cobalt additions increased the sugar level in blood in one case,[154] but in another they lowered the blood sugar level while augmenting the liver glycogen content.[209] In sheep, the administration of cobalt increased the absorption of glucose in the small intestine.[210] Daily additions of 20–60 mg cobalt chloride to cattle increased the carotene content of their blood,[155,211,212] but a supplement of 4 mg to sheep did not alter carotene levels.[98]

The addition of 0·1–0·5 mg cobalt chloride kg^{-1} live weight increased the secretion of bile-pancreatic juice in sheep, but a dose of 1 mg depressed secretory activity.[213-215] The introduction of varying quantities of cobalt into sheep rations increased both gastric secretion and saliva.[216,217]

Small additions of cobalt salts have been reported to improve the reproductive capacity of cattle and sheep.[218,219] Augmenting feedstuffs with cobalt has been found to decrease acetone bodies in the urine of cows;[220] irrespective of whether or not thiabendazole was given, cobalt-deficient sheep excreted considerably greater amounts of methylmalonic acid in the urine than did cobalt-treated animals.[221] The addition of about 0·5 mg cobalt chloride per day for varying periods stimulated the lactation of ewes.[222,223]

Administration of cobalt compounds has been observed to increase enzyme activity in ruminants. Lipase, amylase, diastase, trypsin,[213-215] and nuclease[224] in sheep, and arginase activity in cattle,[225] were improved by cobalt supplements.

One worker reported an increased resistance to tuberculosis induced by cobalt additions in cattle; this was attributed to the inhibitory action of cobalt on the formation and execretion of ketone bodies.[226]

Cobalt dosing had no effect on selenium concentration in the liver of sheep; cobalt-deficient animals accumulated more selenium in their kidneys than did healthy sheep.[337]

A daily supplement of 20 mg cobalt chloride to cows produced more fat, casein, lactose, ash, dry matter, and trace elements in milk.[136] In sheep, a study of the flow of cobalt along the digestive tract indicated a significant net secretion of cobalt in the stomach, and a net absorption of this element from both the small and large intestine.[228]

Chickens

For a long period, cobalt was only considered indispensible for cattle and sheep, but in recent years supplements of this element to the rations of other animals and birds have often proved beneficial. The role of cobalt in the metabolism of chickens will be surveyed in this section.

Many papers have shown that the addition of small quantities of cobalt compounds to poultry rations has improved the growth and development of chickens.[229–239] Additions of cobalt to the feed, in mg kg^{-1} of dry feed, have varied appreciably: 0·02–0·2 cobalt chloride over 40 days,[231] 1·2 cobalt,[232] 0·1 cobalt,[233] 0·62–62 cobalt chloride and cobalt carbonate,[234] 0·1–2 cobalt chloride,[235] 0·796 and 1·178 cobalt,[236,237] 1 cobalt.[239] One investigator was unable to find a benefit to either growth rate or development with additions of 1·2–5·9 mg cobalt kg^{-1} of feed.[240]

Workers have stated that cobalt stimulated the bacterial synthesis of vitamin B_{12} in the intestines of chicks,[241] and that the element is nutritionally important for chicks on diets which are lacking in choline and vitamin B_{12}.[242] Others express the opinion that cobalt is apparently used by the hens for the biosynthesis of vitamin B_{12}, but the addition of a cobalt compound to the diet does not guarantee quantities of B_{12} required for satisfactory hatchability and sufficient transfer to hatched chicks.[243]

In a study of the metabolism of cobalt and of vitamin B_{12} in the hen, it was found that after administration of radioactive cobalt chloride, only a very small amount of the cobalt was still in the inorganic state; the cobalt complexes formed were mainly not cobalamins.[244] The hatchability of hen eggs immersed for 10 min in 1–10% cobalt chloride solution was markedly diminished.[245] Cobalt additions to the rations of White Leghorns lowered the mortality of cocks and hens.[246]

A number of workers have studied various aspects of cobalt in the biochemistry and physiology of chickens. A daily addition of 0·25 mg cobalt sulphate per chick increased the concentration of hemoglobin and the number of leucocytes.[247-249] When radioactive cobalt chloride was administered orally, 50% was found in the cecum, available for corrinoid synthesis.[250] The total utilization of cobalt for the synthesis of corrinoids by microorganisms increased with the length of incubation period and the age of the chickens.[251]

The addition of 0·79 mg cobalt kg⁻¹ of feed increased the iron and cobalt content of blood and bone marrow.[252,253] When 0·1–2 mg of cobalt chloride was added to the diet of young chickens, a significant increase of total nitrogen in blood serum was observed.[254] When phosphorus additives were incorporated into the chicken rations, cobalt concentration in the liver gradually decreased for the first 50 days but increased again after 65 days.[255] Hens given cobalt chloride in their feed showed an increase in blood transaminases.[256] The addition of 0·1–2 mg cobalt chloride kg⁻¹ live weight to chicks 30 days before immunization against *Salmonella pullorum* increased the intensity of immunization and the synthesis of agglutinins.[257]

The average cobalt content of blood in chickens has been placed at 0·045 mg kg⁻¹, and that of liver at 0·12 mg kg⁻¹.[258] The average cobalt in shelled eggs has been reported as 0·03 mg kg⁻¹;[259] uncooked chicken meat contained 0·0057 mg cobalt kg⁻¹.[260]

Oyster shells have been found to contain about 1 mg cobalt kg⁻¹, and it has been suggested that they would be beneficial in poultry nutrition.[261]

Pigs

Some years elapsed between the discovery of the role of cobalt in ruminant nutrition and the realization that other animals, in some areas and under certain conditions, can benefit from the addition of cobalt to their rations. A number of investigators have reported weight gains and an improvement in the general development of pigs from cobalt additions.[62,262-276] Unfortunately, in the literature the cobalt additions have been expressed in different ways: mg kg⁻¹ of feed, mg kg⁻¹ of live weight of the pig, or mg per animal per day. Accordingly, comparisons of all results are not readily made.

Increased weight gains have been reported for the following additions, in mg kg⁻¹ of feed: 0·075–0·15 cobalt,[268] 1·97 cobalt,[263,264] 1·84–2·1 cobalt,[269] 3–6 mg cobalt chloride,[270] and 3·8 mg cobalt chloride.[271] Benefits have been indicated when cobalt has been added, in mg kg⁻¹ of body weight: 0·1 cobalt chloride,[272] 0·1–0·2 cobalt,[273] and 0·5 cobalt chloride.[274] Improved results have been obtained by cobalt supplements, in mg per pig per day, by 4–10,[275] 10,[262] and 400–600.[276]

On the other hand, three papers present data which indicate that the following additions of cobalt chloride did not improve weight gain:

0·3 mg kg^{-1} body weight,[277] 2 mg kg^{-1} dry feed,[278] and 2·47–11·59 mg per head per day.[279]

Cobalt additions of 0·03–0·8 mg kg^{-1} body weight have been found to increase the hemoglobin content in pigs.[280-282] Cobalt supplements increased the concentration of cobalt mainly in the liver and kidneys,[273] and had a positive effect on nitrogen metabolism.[283] The addition of 2 mg cobalt kg^{-1} of feed to a low-protein diet increased the fat content of pigs.[284] Cobalt supplementation was associated with an increased ratio of heart weight to body weight.[285]

It has been reported that cobalt supplements stimulated the reproductive capacity of pigs,[218] and 0·97 mg of cobalt kg^{-1} dry weight of fodder has been recommended to assist pigs in adapting to increased temperatures.[286] In a study of human and animal myocardia, the highest cobalt was found in pigs, at a level of 0·046 mg kg^{-1}.[287] A food balance in pigs showed an intake in fodder and water of 0·1333 mg, excretion in feces and urine of 0·1059 mg, leaving a balance in the body of 0·0274 mg.[288] The cobalt content of pig liver, on the dry basis, was found to be 1·18 mg kg^{-1} for newborn pigs and 0·42 for 7-month-old pigs.[289] Iodine deficiency markedly depresses the content of cobalt in sows' milk.[290]

Rabbits

A considerable amount of work has been carried out on the effect of cobalt on various aspects of rabbit nutrition; most of these investigations arise from the common use of the rabbit as an experimental animal in biochemical and physiological studies.

Additions of cobalt to rations in quantities ranging from 0·07–0·1 mg cobalt kg^{-1} of feed, and 0·1–0·25 mg cobalt sulphate kg^{-1} of feed, resulted in increased weight gains.[291-293] Weight gains were also recorded when rabbits were fed diets containing vegetables rich in cobalt.[294]

Cobalt supplements of 0·1–0·25 mg cobalt sulphate kg^{-1} of feed, or 0·7–1 mg cobalt chloride per day, increased plasma protein levels.[292,293,295] The addition of cobalt gave a higher content of vitamin B_{12} in liver tissue;[296] conversely, the vitamin B_{12} level fell when rabbits were fed cobalt-deficient rations.[297] It was found that the content of cobalt rose in the livers of rabbits fed a vitamin E-deficient diet on the 11th to 17th day.[298] The inhibition of thrombocyte aggregation observed in rabbits after parenteral administration of cobalt was due to alterations in the fibrinogen localized on the outer cell membrane.[299]

In the mechanism of the stimulating effect produced by cobalt on erythropoiesis, an important role is played by tissue hypoxia.[300]

Subcutaneous administration of cobalt chloride caused liver steatosis, or fatty degeneration, in rabbits.[301] Cobalt is higher in aorta, liver, kidneys, and adrenal glands than in other organs of healthy rabbits.[302] The addition of 0·25 mg cobalt nitrate kg^{-1} body weight produced decreases in the water and phosphorus content in the femoral bones, and in the calcium content of liver.[303]

Rabbits on a high-cholesterol diet given 6 mg cobalt kg^{-1} were found to have increased serum cholesterol and atherosclerotic changes in the aorta.[304] The administration of 0·1–5 mg cobalt sulphate kg^{-1} live weight increased the cobalt content in rabbit blood and muscles.[305] Cobalt sulphate at 0·05–2 mg kg^{-1} weight lowered the content of neuraminic acid in rabbits, and caused an increase at 5 mg kg^{-1} weight.[306] In cobalt-treated rabbits, alterations of the primary structure caused changes in the conformation of fibrinogen, resulting in hemorrhagic diathesis.[307] When 0·25 mg cobalt chloride was given daily to rabbits, there was no change observed in the physicochemical properties of blood; 25 mg daily for 41 days, or 50 mg for 14 days, increased blood viscosity, reduced electrical conductivity, and increased the osmotic pressure, resulting in death to the animals.[308] In a study of the effect of cobalt on intraocular tension in rabbits, the administration of cobalt resulted in a decrease of coefficient of outflow and a minute increase in volume of the eye chamber.[309]

Rats and Mice

The extensive literature on the effect of cobalt on rats and mice reflects, of course, the importance of these animals in laboratory investigations concerned with many aspects of biochemistry, physiology, pharmacology, nutrition, microbiology, environmental protection, and other life sciences.

At 500 mg cobalt kg^{-1} of rations, cobalt is growth-depressing; decreased excretion of both ascorbic and glucuronic acids in urine was observed.[310] Following the injection of cobalt chloride, the main organs which accumulated cobalt were the liver and kidneys; a large part of the injected dose was excreted rapidly.[311] About 27% of cobalt-60 administered to rats was absorbed by the tissues; after 2 months only about 4% remained in the organism.[312] The cobalt content in the brain of rats is the same at all ages; in the liver it decreases up to the twelfth month and then increases to the original value, and in muscle the cobalt levels oscillate.[313]

Several papers have recorded observations on hemoglobin and other aspects of blood in rats.[314–318] One worker found that large doses of cobalt increased hematin synthesis,[315,316] whereas another observed many toxic symptoms when the cobalt addition exceeded 1·25 mg kg^{-1} of feed.[317] The parenteral administrationn of cobalt compounds disturbed blood clotting in rats.[318]

The effect of cobalt on the thyroid gland has been studied.[319-321] When supplies of iodine were low, 0·04–0·4 mg cobalt per rat per day gave higher weights of thyroid, but over 0·4 mg tended to suppress the thyroid function.

Cobalt may inhibit the biosynthesis of lipoproteins in liver.[322] Except for histidine, tryptophan, and cysteine, amino acids did not influence the accumulation of cobalt in rat liver.[323] Cobalt is without effect against liver necrosis in the rat.[324]

Cultures of rat dermal fibroblasts will not normally tolerate cobalt levels above 2·5 μg ml^{-1}, but repeated intermittent treatment with cobalt chloride produced fibroblasts able to tolerate 7·5 μg cobalt ml^{-1}.[325] The cobalt-60 EDTA chelate is partly bound to skin proteins.[326] The addition of 1–64 μg cobalt chloride ml^{-1} to a culture of mouse fetus fibroblasts inhibited proliferation.[327] The introduction of powdered cobaltous oxide into the lungs of rats produced a more rapid and lethal edema than did that of cobaltic oxide.[328] Cobalt chloride in aerosol form caused acute lung edema.[329]

Some studies have been made on the absorption of cobalt.[330-333] When cobalt was administered in milk, absorption from the gastrointestinal tract was 40%.[330] Uptake of cobalt in endocrine organs was greatest from 2–4 h after injection.[331] The addition of cobalt diminished the mucosal uptake of iron from the intestinal lumen;[332] in anemic mice, a close relation was found between the absorptive pathways of iron and cobalt.[333]

The effect of cobalt on polyneuritic rats was unfavorable, and raised their need for vitamin B_1.[334] Administration of 0·01 M cobalt chloride to mice resulted in severe depression of γ-globulins.[335] Rats on a cobalt-deficient diet had a higher incidence and intensity of dental caries.[336] It appears that cobalt not only abolishes the metabolic response of rats to cold, but also reduces their basal oxygen consumption.[337] Following partial removal of the spleen in rats, the rest of the spleen grew stronger in animals treated with cobalt.[338] Cobalt implantation in the brain is known to make most laboratory animals epileptic; homolateral carotid artery ligation and ischemia in rats alleviates this epileptic state.[339]

The addition of cobalt increased the content and reactivity of free sulphhydryl groups in the central nervous system of rats.[340] The injection of cobalt chloride into rats depressed liver respiration.[341] Dihydrolipoic acid readily forms a chelate with cobalt, which is oxidized immediately in air, and the biological activity of the co-enzyme is lost.[342] Low doses of cobalt chloride, 1–5 mg kg^{-1} intravenously (i.v.), result in a stimulation of respiration; higher doses of 25–30 mg kg^{-1} are lethal, producing central respiratory paralysis and heart injuries.[343] Rats injected with cobalt showed a marked inhibition of oxygen uptake in the liver and medullar of the kidneys.[344] Cobalt treatment reduced the ferricyanide-linked ketoglutarate dehydrogenase activity of isolated rat heart mitochondria.[345]

Other Mammals

A few publications have considered the role of cobalt in the growth and development of animals other than those already reviewed in this chapter. We shall examine the recorded information on the effect of cobalt on the beaver, buffalo, dog, fox, guinea pig, hamster, horse, mink, and reindeer.

Perhaps the most intriguing reference to the effect of cobalt in animal nutrition is one relating to willow bark in beaver lands along the Dnieper River. In the upper section of this river, where no beavers are found, the bark of willows, which is the principal food of these animals, is very low in cobalt, 0.066 mg kg^{-1} of dry weight. The authors concluded that cobalt deficiency was the factor which restrained the beavers from settling in this area.[346] This opinion may be wiser than the facts might warrant, for the beaver is not only a very industrious animal but also a remarkably intelligent one.

Several publications have considered the role of cobalt in buffaloes. Additions of 0.1 mg cobalt chloride kg^{-1} of feed increased milk yield, fat, and the albumin:globulin ratio of blood.[347] Cobalt supplements increased the catalase activity of buffalo blood;[348] 0.1–0.3 mg cobalt chloride kg^{-1} live weight increased hemoglobin, erythrocyte, and leucocyte counts, and also body weight.[349] Cobalt was found, in mg kg^{-1}, at levels of 0.045–0.048 in buffalo liver, 0.035–0.042 in muscle, 0.012–0.017 in blood, and 0.007 in milk.[350]

Several observations have been made on cobalt in dogs used as experimental animals. A subcutaneous injection of 2.55 mg kg^{-1} weight lowered the concentrations of blood tyrosine and adrenaline.[351] In the liver, spleen, lungs, and muscles of dogs killed by radiation, the cobalt levels were below normal.[352] Administration of small doses of cobalt increased salivation, but large doses repressed it.[353]

The effect of cobalt on the growth and development of silver and Arctic foxes has been documented. Additions of 0.3–0.5 mg cobalt chloride kg^{-1} body weight to the rations of silver foxes stimulated reproductive capacity.[218] A supplement of 0.5 mg cobalt chloride kg^{-1} body weight increased the hemoglobin content but had no effect on protein levels of blood serum;[354] the same addition resulted in a substantial increase of the cobalt balance in the body.[355] When the ration of young foxes was augmented with 0.5 μg cobalt chloride kg^{-1}, favorable effects were noted on body-weight gains, body length, chest measurements, and fur quality.[356] The cobalt level in foxes has been tabulated as follows, (in mg kg^{-1}: blood, 0.046–0.052; heart, 0.035–

0·114; kidneys, 0·055–0·081; liver, 0·110–0·146; muscles, 0·041–0·051; pancreas, 0·173–0·181; and spleen, 0·113–0·146.[357]

The administration of cobalt compounds to guinea pigs has been described by a few investigators. Cobalt chloride exerted a hypertensive action on the anesthetized guinea pig.[358] In scorbutic guinea pigs, cobalt deposition increased in the liver, muscles, and brain tissues, and decreased in the spleen.[359] In another study of animals suffering from scurvy, the addition of cobaltous chloride increased the ascorbic acid content in spleen and adrenal glands by the simultaneous decrease of the vitamin C content in other organs, and did not accelerate the animals' recovery.[360] Administration of 30 mg cobalt kg^{-1} body weight per day for 6–7 days produced necrosis of the liver, disappearance of glycogen and appearance of lipids in the injured cells;[361] the same worker found the addition of 80 mg cobalt kg^{-1} body weight had a very detrimental effect on guinea pigs' kidneys.[362] Administering 0·06–0·10 M cobalt chloride to an isolated guinea pig heart perfused in vitro decreased both mechanical performance and substrate uptake.[363]

A study of cobalt balance in the hamster showed that 10–14 % was excreted in the urine, and 86–90 % in the feces.[364] Small additions of cobalt have been observed to decrease cholesterol in liver and blood serum, liver fatty acids and phospholipids.[365,366]

It has been reported that the average cobalt content of kumiss, fermented mare's milk, is 0·07 mg kg^{-1}; it contains more cobalt than cow milk.[367] Horses fed a cobalt supplement showed increased contents of albumin and hemoglobin in blood, and of protein in blood plasma.[368]

When young mink were fed 0·5 mg cobalt chloride kg^{-1} body weight, the weight gains decreased, hemoglobin increased, and no effect was found on protein of blood serum.[354] A cobalt balance on mink having the same cobalt-containing ration showed a substantial increase in the body cobalt.[355] The cobalt content in mink has been reported as follows, in mg kg^{-1}: blood, 0·076; heart, 0·146; kidneys, 0·106–0·153; liver, 0·141; muscles, 0·064; pancreas, 0·225; and spleen 0·259–0·425.[357] One investigator found that the addition of 0·25 mg cobalt chloride kg^{-1} live weight improved the growth rate, blood indices, and reproduction of mink.[369]

A depletion of cobalt in reindeer in the winter has been described,[370] and the addition of this element is recommended for winter and spring feeding of reindeer.[371]

Other Birds

The effect of cobalt on the nutrition of chickens has already been summarized in this chapter; its influence on geese, pheasants, and turkeys has been the subject of a few publications. A daily addition of 1 mg cobalt choride kg^{-1} body weight to geese rations enhanced the rate of growth, increased blood hemoglobin, improved egg-laying, and increased the number of young geese.[372,373] Another worker found that daily feeding of 0·5 mg cobalt chloride per goose increased egg production, hatchability, and weight of the hatched goslings; a dose of 1 mg was less effective.[374] After a supplement of 1 mg cobalt kg^{-1} live weight for 60–136 days, the cobalt content, expressed as a percentage gain over that in control birds, was approximately: blood, 162; egg yolk, 54; feathers, 45; liver, 246; and muscles, 720.[375]

In a study of cobalt levels in foods and livers of pheasants, the cobalt content of pheasant livers was 0·23 mg kg^{-1}, and those of corn, rosehips, foxtail seeds, and ragweed seeds were respectively, in mg kg^{-1}, 0·040, 0·069, 0·084, and 0·228. It was concluded that cobalt is not a limiting nutritional factor for pheasants in northern Minnesota.[376]

A publication places the requirement of turkeys in south-eastern Kazakhstan at 0·8 mg cobalt kg^{-1} of dry feed.[377]

Fish and Shellfish

A few papers have presented limited data on the cobalt content of fish. In three types of shellfish, *Meretrix meretrix*, *Paphia philippinarum*, and *Corbicula leana*, the cobalt in the shell was 0·03–0·04 mg kg^{-1}.[378] The cobalt contents of ocean fish, freshwater fish, mollusks, and crustaceans, respectively, have been given, in mg kg^{-1}, as 0·1, 0·09, 0·01, and 0·04.[379] Figures have been published for cobalt levels, in mg kg^{-1}, in cod muscle, 0·0021; cod liver, 0·0111; and carp muscle, <0·001.[380] Bonito heart was found to contain up to 0·14 mg cobalt kg^{-1}.[381]

The addition of 0·4 mg cobalt nitrate kg^{-1} of carp led to increases of 4 % of erythrocytes and 13 % of hemoglobin in blood, of 16 % in brood yield, and of 15·4 % in total fish productivity.[382]

Insects

Investigations on the role of cobalt in the growth and metabolism of insects have been almost entirely restricted to bees and silkworms. The addition of cobalt, at the rate of 1 mg l^{-1} of syrup, to bee colonies increased the honey yield and brood rearing.[383] Another paper reported that the addition of 0·47 mg cobalt chloride l^{-1} of a syrup of sugar and water was most effective, increasing both honey and wax production, and rearing of the brood.[384] In the Odessa district, where the soil has average cobalt levels of 1·52–2·44 mg kg^{-1}, the addition of 4–16 mg cobalt l^{-1} of sugar syrup increased

honey production; 4 mg l⁻¹ is recommended.[385] In Russia, a study of trace
elements in honey and plant pollen indicated that cobalt was the least
abundant in pollen and especially in honey; it was recommended that feed for
bees should be supplemented with cobalt.[386]

In Russia, the addition of cobalt chloride was found to improve the
productivity of silkworms by enhancing the positive balance of protein, live
weight of caterpillar, cocoon, and silk envelope.[387] Another investigator
found that cobalt did not increase the weight of the cocoon, but the weight and
percentage of silk capsule was higher.[388] Spraying mulberry trees with either
the chloride or sulphate of cobalt increased the weight of silkworms, and the
weight of their cocoons.[389] Spraying twice daily with a solution containing
100 mg cobalt l⁻¹ at the rate of 0·1 ml per leaf, increased the growth rate of
silkworms in India; cobalt became toxic at 1 g l⁻¹.[390]

The larvae of the insect *Plodia interpunctella* reared on an artificial medium
were found to have a tolerance for cobalt of 3·5, where the tolerance for
arsenic is taken as 1.[391]

Toxicity

The possibility of livestock poisoning caused by an excess of cobalt in natural
herbage or feedstuffs is extremely remote. Many years ago it was found that
growing cattle can consume up to 50 mg cobalt per 45 kg body weight without
ill effects,[73] and it was demonstrated that sheep can tolerate up to 160 mg
daily per 45 kg weight for at least 8 weeks without harmful results, but that
higher dosages are injurious.[392] A later experiment in New Zealand reported
that all five ewes involved died from a single dose of 3 oz of $CoSO_4.7H_2O$,[393]
but it must be realized that this represents about 3·5 g cobalt per ewe, a
quantity far in excess of normal supplements, and virtually unobtainable except
through human error. A paper from Australia, observing that symptoms of
cobalt toxicity in cattle are not specific and, in fact, are similar to those of
cobalt deficiency, concluded from a high cobalt level in the liver that cobalt
toxicity seemed evident in five field cases.[394]

Some work has been done on the inhalation of cobalt-containing dust by
animals. Inhalation of 2–160 μg cobaltous oxide l⁻¹ by hamsters resulted in
less than 1 % retention in the lungs, and none after 6 days.[395] Chronic
inhalation of cobalt dust at the threshold limit value of 0·1 mg m⁻³ did not
adversely affect lung function in animals.[396] When miniature swine inhaled
cobalt metal powder at 0·1–1·0 mg m⁻³, lung disease was detected.[397]

On soil heavily contaminated with industrial waste, cabbages had a content
of 0·69 mg cobalt kg⁻¹ and carrots contained 0·88 mg cobalt kg⁻¹, but these
had no detrimental effect when fed to rabbits.[398]

References

1. Aston, B. C. (1932). *New Zealand J. Agric.* **44**, 367–78.
2. Askew, H. O. (1946). *New Zealand J. Sci. Technol.* **28A**, 37–43.
3. Rigg, T. (1950). *Research* **3**, 131–5.
4. Kidson, E. B. (1937). *New Zealand J. Sci. Technol.* **18**, 694–707.
5. Kidson, E. B. (1938). *J. Soc. Chem. Ind.* **57**, 95–6.
6. Dixon, J. K. (1937). *New Zealand J. Sci. Technol.* **18**, 892–7.
7. McNaught, K. J. (1938). *New Zealand J. Sci. Technol.* **20A**, 14–30.
8. McNaught, K. J. (1948). *New Zealand J. Sci. Technol.* **30A**, 26–43.
9. Filmer, J. F. and Underwood, E. J. (1937). *Aust. Vet. J.* **13**, 57–64.
10. Underwood, E. J. and Harvey, R. J. (1938). *Aust. Vet. J.* **14**, 183–9.
11. Underwood, E. J. (1971). "Trace Elements in Human and Animal Nutrition", 3rd Edn. Academic Press, New York and London.
12. Marston, H. R. and Lee, H. J. (1952). *Nature, Lond.* **170**, 791.
13. Marston, H. R. and Smith, R. M. (1952). *Nature, Lond.* **170**, 792–3.
14. Beeson, K. C. (1950). U.S. Dept. Agric. Inf. Bull. 7.
15. Stiles, W. (1961). "Trace Elements in Plants", 3rd Edn. Cambridge University Press, London.
16. Young, R. S. (1956). *Sci. Prog.* **44**, 16–37.
17. Cunningham, I. J. (1955). *Adv. Vet. Sci.* **2**, 138–81.
18. Underwood, E. J. (1959). *Ann. Rev. Biochem.* **28**, 499–526.
19. Young, R. S. (1960). "Cobalt", Am. Chem. Soc. Monograph 149. Reinhold, New York.
20. de Vuyst, A., Arnould, R., Vanbelle, M., Vervack, W., and Moreels, A. (1960). *Agricultura* **8**, 509–51.
21. Mills, C. F. (1970). "Trace Element Metabolism in Animals". Livingstone, London.
22. Ammerman, C. B. (1970). *J. Dairy Sci.* **53**, 1097–1107.
23. Masoero, P. (1972). *Atti Simp. Int. Agrochim.* **9**, 539–81; (1973). *C.A.* **79**, 77221.
24. Hoekstra, W. G., Suttie, J. W., Ganther, H. E., and Mertz, W., eds (1974). "Trace Element Metabolism in Animals. 2. Proceedings of the Second International Symposium". University Park Press, Baltimore.
25. Lee, H. J. (1975). *In* "Trace Elements in Soil–Plant–Animal Systems" (D. J. D. Nicholas and A. R. Egan, eds) 39–54. Academic Press, New York and London.
26. Underwood, E. J. (1975). *I* "Trace Elements in Soil–Plant–Animal Systems" (D. J. D. Nicholas and A. R. Egan, eds) 227–41. Academic Press, New York and London.
27. Smith, R. M. and Gawthorne, J. M. (1975). *In* "Trace Elements in Soil–Plant–Animal Systems" (D. J. D. Nicholas and E. R. Egan, eds) 243–58. Academic Press, New York and London.
28. Suttle, N. F. (1975). *In* "Trace Elements in Soil–Plant–Animal Systems" 271–89. Academic Press, New York and London.
29. Egan, A. R. (1975). *In* "Trace Elements in Soil–Plant–Animal Systems' (D. J. D. Nicholas and E. R. Egan, eds) 371–84. Academic Press, New York and London.
30. Underwood, E. J. (1975). *Nutr. Rev.* **33**, 65–9.
31. Mitchell, R. L. (1947). *Research* **1**, 159–65.
32. Pickett, E. E. (1960). Missouri Agric. Exp. Stn Res. Bull. 724.
33. Price, N. O. and Hardison, W. A. (1963). Virginia Agric. Exp. Stn Bull. 165.
34. Price, N. O. and Huber, J. T. (1964). Virginia Agric. Exp. Stn Bull. 177.
35. Kubota, J. (1964). *Soil Sci. Soc. Am. Proc.* **28**, 246–51.

36. Ignat'ev, V. N. and Levin, M. M. (1968). *Trudȳ Mold. Nauch.-Issled. Inst. Zhivotnovod. Vet.* **4**, 170–2; (1972). *C.A.* **77**, 138538.

37. Pereira, J. A. A., Da Silva, D. J., and Braga, J. M. (1971). *Experientia* **12**, 155–88; (1972). *C.A.* **76**, 125761.

38. Rampilova, M. A., Petrovich, P. I., Belokurova, E. I., Emedeeva, N. I., and Kharitonov, Yu. D. (1974). *Mikroelem. Sib.* **9**, 114–18; (1975). *C.A.* **83**, 77227.

39. van der Merwe, F. J. (1959). *S. African J. Agric. Sci.* **2**, 141–63.

40. Kondratenko, I. P. (1972). *Nauch. Tr., Sev.-Zapad. Nauch.-Issled. Inst. Sel'. Khoz.* No. 22, 39–41; (1973). *C.A.* **79**, 17202.

41. Kudashev, A. K. (1972). *Khim. Sel'. Khoz.* **10**, 701–2; (1973). *C.A.* **78**, 28275.

42. Kummer, H., Von Polheim, P., and Scholl, W. (1973). *Landw. Forsch., Sonderh.* **28**, 215–27; (1974). *C.A.* **80**, 81417.

43. Chodan, J. (1962). *Rocz. Nauk Roln., Ser. F.* **75**, 545–62; (1963). *C.A.* **59**, 15889.

44. Uzilovskaya, P. Sh., Uspenskaya, M. V., and Ivoshina, V. I. (1962). *Trudȳ Nauch.-Issled. Inst. Zhivotnovod. Uzbek. Akad. Sel'.-khoz. Nauk* No. 7, 27–31; (1963). *C.A.* **59**, 5559.

45. Korovin, N. K. (1964). Mikroelem. Biosfere Ikh Primen. Sel'. Khoz. Med. Sib. Dal'nego Vostoka, Dokl. Sib. Konf. 2nd, 464–6; (1969). *C.A.* **70**, 65709.

46. Gustun, M. I. (1965). *Khim. Sel'. Khoz.* **3**, 67–8; (1966). *C.A.* **64**, 7311.

47. Mel'nikova, M. F. and Glushkova, N. A. (1966). *Trudȳ Kirov. Sel'.-khoz. Inst.* **19**, 22–6; (1968). *C.A.* **68**, 10755.

48. Yudkin, F. E. (1968). Sb. Mater. Zon. Nauch.-Proizvod. Konf. Nauch. Prakt. Vet. Rab. Urala, Sib. Dal'nego Vostoka, 201–5; (1972). *C.A.* **76**, 98256.

49. Makmudov, Kh. Kh. (1972). *Vest. Sel'.-khoz. Nauki* **15**, 119–22; (1973). *C.A.* **78**, 83234.

50. Zusmanovskii, A. G., Magnitskii, P. V., Merkulov, N. N., and Kornev, S. D. (1972). *Trudȳ Ul'yanovsk. Sel'.-khoz. Inst.* **17**, 108–18; (1973). *C.A.* **79**, 41333.

51. Scharrer, K. and Judel, G. K. (1959). *Z. Tierphysiol. Tierernähr. Futtermittelk.* **14**, 34–42; (1959). *C.A.* **53**, 18329.

52. Narevicius, J. (1961). *Trudȳ Litovsk. Vet. Akad.* **6**, 27–53; (1965). *C.A.* **62**, 8178.

53. Walsh, T., Fleming, G. A., Kavanagh, T. J., and Ryan, P. (1955–56). *Eire Dept. Agric. J.* **52**, 56–116; (1959). *C.A.* **53**, 6500.

54. Kazaryan, E. S. and Airuni, G. A. (1967). *Izv. Sel'.-khoz. Nauk, Min. Sel'. Khoz. Arm. S.S.R.* **10**, 65–72; (1968). *C.A.* **68**, 28789.

55. Russel, A. J. F., Whitelaw, A., Moberly, P., and Fawcett, A. R. (1975). *Vet. Rec.* **96**, 194–8.

56. Mitchell, R. L. (1972). *Atti Simp. Int. Agrochim.* **9**, 521–32.

57. Kazaryan, E. S., Asratyan, G. S., and Stepanyan, M. S. (1965). *Soobshch. Lab. Agrokhim., Akad. Nauk Arm. S.S.R.* No. 6, 20–30; (1966). *C.A.* **65**, 6236.

58. Poole, D. B. R., Moore, L., Finch, T. F., Gardiner, M. J., and Fleming, G. A. (1974). *Ir. J. Agric. Res.* **13**, 119–22.

59. Gardiner, M. R. and Nairn, M. E. (1969). *Aust. Vet. J.* **45**, 215–22.

60. Bennetts, H. W. (1959). *J. Dept. Agric. W. Aust.* **8**, 631–6, 639–48.

61. Morozov, S. D. and Savchitskaya, S. S. (1955). Mikroelementy v Sel'. Khoz. i Med., Akad. Nauk Latv. S.S.R., Otdel. Biol. Nauk, Tr. Vses. Soveshch, Riga, 603–5; (1959). *C.A.* **53**, 11561.

62. Roslyakov, I. A. K. and Baiturin, M. A. (1960). *Trudȳ Alma-Atinsk. Zoovet. Inst.* **12**, 148–52; (1963). *C.A.* **59**, 4318.

63. Valuiskii, P. P. and Odynets, R. N. (1968). Mikroelem. Zhivotnovod. Rastenievod., 76–82; (1970). *C.A.* **72**, 108470.

64. Barkan, Ya. G. (1972). Mikroelem. Biosfere Ikh Primen. Sel'. Khoz. Med. Sib. Dal'nego Vostoka, Dokl. Sib. Konf. 4th, 40–7; (1975). *C.A.* **83**, 74012.

65. Drebickas, V. (1961). *Liet. Gyvulininkystes Mokslų Tyrimo Inst. Darb.* **5**, 125–39; (1963). *C.A.* **59**, 2010.
66. Melkumyan, R. S. and Dadayan, A. Kh. (1973). *Zhivotnovodstvo* No. 11, 52–3; (1974). *C.A.* **80**, 69497.
67. Levshin, D. N. (1962). *Uchen. Zap. Kazansk. Vet. Inst.* **88**, 89–98; (1964). *C.A.* **61**, 15099.
68. Sirotkin, V. I. and Sushko, T. A. (1968). *Trudy̅ Primorsk. Sel'.-khoz. Inst.* No. 3, 9–11; (1973). *C.A.* **78**, 14666.
69. Yanchilin, L. V., Alekseev, M. I., and Vlasenko, N. I. (1972). *Sib. Vest. Sel'.-khoz. Nauki* **2**, 70–6; (1973). *C.A.* **78**, 146534.
70. Burenbayar, R. (1972). Mikroelem. Biosfere Ikh Primen. Sel'. Khoz. Med. Sib. Dal'nego Vostoka, Dokl. Sib. Konf. 4th, 291–4; (1975). *C.A.* **82**, 110621.
71. Odynets, R. N. (1962). Mikroelementy v Zhivotnovod., Min. Sel'. Khoz. S.S.S.R., Vses. Akad. Sel'.-khoz. Nauk Otdel. Zhivotnovod., 44–52; (1963). *C.A.* **58**, 11737.
72. Pytloun, J., Alexa, Z., Markovic, P., and Plickova, U. (1974). *Biol. Chem. Vyz. Zuirat* **10**, 369–78; (1975). *C.A.* **82**, 169168.
73. Keener, H. A., Percival, G. P., and Morrow, K. S. (1949). *J. Dairy Sci.* **32**, 527–33.
74. Lapin, L. N. (1968). Mater. Respub. Konf. Probl. "Mikroelem. Med. Zhivotnovod." 1st, 137–8; (1971). *C.A.* **74**, 29481.
75. Chubinskaya, A. A. (1962). Mikroelementy v Zhivotnovod., Min. Sel'. Khoz. S.S.S.R., Vses. Akad. Sel'.-khoz. Nauk Otdel. Zhivotnovod., 107–13; (1963). *C.A.* **58**, 11738.
76. Kossila, V. (1976). *Karjantuote* **59**, 11; (1976). *C.A.* **85**, 107722.
77. Donaldson, L. E., Harvey, J. M., Beattie, A. W., Alexander, G. I., and Burns, M. A. (1964). *Queensland J. Agric. Sci.* **21**, 167–79.
78. Popov, I. S. and Marnov, D. I. (1959). *Izv. Timiryazev. Sel'.-khoz. Akad.* No. 5, 123–38; (1960). *C.A.* **54**, 25124.
79. Stewart, J., *et al.* (1955). *Vet. Record* **67**, 755–6; (1959). *C.A.* **53**, 13301.
80. Odynets, R. N. and Aituganov, M. D. (1969). *Izv. Akad. Nauk Kirg. S.S.R.* No. 4, 48–58; (1970). *C.A.* **72**, 118938.
81. Lee, H. J. and Marston, H. R. (1969). *Aust. J. Agric. Res.* **20**, 905–18.
82. Gaffarov, A. K. and Kamolov, A. (1974). *Trudy̅ Tadzh. Sel'.-khoz. Inst.* **19**, 66–78; (1976). *C.A.* **84**, 88330.
83. Odynets, R. N. (1966). Mikroelem. Sel'. Khoz. Med., Dokl. Vses. Soveshch. Mikroelem. 5th, 507–13; (1970). *C.A.* **73**, 1321.
84. Stepanova, S. A. and Toshchev, V. K. (1974). *Trudy̅ Yarosl. Nauch-Issled. Inst. Zhivotnovod. Kormaprozvod.* **4**, 97–106; (1976). *C.A.* **84**, 29494.
85. Kolesov, A, M., Zamarin, L. G., Emel'yanov, A. N., Tarasov, I. I., Tabakova, N. I., Tkacheva, K. I., and Zlotina, V. S. (1961). *Trudy̅ Mosk. Vet. Akad.* **37**, 114–16; (1962). *C.A.* **57**, 5087.
86. Dadashev, Ch. N. and Mamedov, R. S. (1967). Mater. Nauch. Konf., Azerb. Nauch.-Issled. Inst. Zhivotnovod., 59–62; (1969). *C.A.* **71**, 121038.
87. Norton, B. W. and Hales, J. W. (1976). *Proc. Aust. Soc. Anim. Prod.* **11**, 393–6.
88. Jones, O. H. and Anthony, W. B. (1970). *J. Anim. Sci.* **31**, 440–3.
89. Dymko, E. F. (1961). *Trudy̅ Alma-Atinsk. Zoovet. Inst.* **12**, 353–74; (1963). *C.A.* **59**, 5559.
90. Odynets, R. N. and Mambetov, M. U. (1960). *Izv. Akad. Nauk Kirg. S.S.R., Ser. Biol. Nauk* **2**, 47–52; (1962). *C.A.* **56**, 3879.

91. Odynets, R. N. and Volostnov, G. A. (1967). *Mikroelem. Zhivotnovod. Rastenievod., Akad. Nauk Kirg. S.S.R.* No. 6, 46–58; (1968). *C.A.* **69**, 50146.
92. Saidov, N. and Gaffarov, A. K. (1974). *Trudȳ Tadzh. Sel'.-khoz. Inst.* **19**, 142–7; (1976). *C.A.* **84**, 88333.
93. Ipatov, P. P. and Vorob'ev, V. I. (1968). *Khim. Sel'. Khoz.* **6**, 775–8; (1969). *C.A.* **70**, 55317.
94. Vorob'ev, V. I. (1968). Mater. Respub. Konf. Probl. "Mikroelem. Med. Zhivotnovod." 1st, 109–10; (1971). *C.A.* **74**, 29469.
95. Roizman, P. Sh. (1955). Mikroelementy v Sel'. Khoz. i Med. Akad. Nauk Latv. S.S.R., Otdel. Biol. Nauk, Trad. Vses. Soveshch., Riga, 589–92; (1959). *C.A.* **53**, 11561.
96. Babin, Ya. A. and Vasyunin, V. V. (1968). *Khim. Sel'. Khoz.* **6**, 861–2; (1969). *C.A.* **70**, 55318.
97. Rudin, V. D. (1962). Mikroelementy i Estestv. Radioaktivn. Pochv. Sb., 227–8; (1963). *C.A.* **59**, 14350.
98. Kazakova, E. M. (1967). *Khim. Sel'. Khoz.* **5**, 52–4; (1967). *C.A.* **66**, 113813.
99. Dergach, V. I., Gorbelik, R. V., |and| Yashchenko, N. F. (1973). *Visn. Sil's'-kogospod. Nauki* No. 7, 96–9; (1973). *C.A.* **79**, 103891.
100. Sidorov, P. I. (1965). *Trudȳ Saratov. Zootek.-Vet. Inst.* **13**, 167–73; (1967). *C.A.* **66**, 83564.
101. Vasyunin, V. V. (1968). Mater. Respub. Konf. Probl. "Mikroelem. Med. Zhivotnovod." 1st, 108–9; (1971). *C.A.* **74**, 29468.
102. Dymko, E. F. (1963). *Trudȳ Alma-Atinsk. Zoovet. Inst.* **13**, 137–42; (1965). *C.A.* **63**, 3384.
103. Pen'kova, A. F. (1966). *Vest. Sel'.-khoz. Nauki* **11**, 69–71; (1967). *C.A.* **66**, 26954.
104. Inoue, R. and Katsukawa, K. (1963). *Hyogo Noka Daigaku Kenkyu Hokoku, Chickusan-gaku Hen* **6**, 37–42; (1966). *C.A.* **64**, 8705.
105. Movsum-Zade, K. K. and Prikhod'ko, M. M. (1975). *Veterinariya* **11**, 74–6; (1976). *C.A.* **84**, 57605.
106. Krasteva, El. (1963). *Nauchni Tr. Vissh. Selskostopanski Inst. "Georgi Dimitrov". Zootek. Fak.* **13**, 109–21; (1964). *C.A.* **61**, 16512.
107. Lesch, S. F. and Penzhorn, E. J. (1964). *S. African J. Agric. Sci.* **7**, 791–5.
108. Hecht, H. (1973). *Archiv. Lebensmittelhyg.* **24**, 255–8; (1974). *C.A.* **80**, 144458.
109. Kirkpatrick, D. C. and Coffin, D. E. (1975). *J. Sci. Food Agric.* **26**, 43–6.
110. Taucins, E. and Svilane, A. (1962). *Trudȳ Sektora Fiziol. Zhivotn., Akad. Nauk Latv. S.S.R., Inst. Biol.* No. 3, 165–205; (1963). *C.A.* **59**, 13174.
111. Berestova, V. I. and Panova, M. K. (1965). *Uchen. Zap. Petrozavodsk. Gos. Univ.* **13**, 107–9; (1966). *C.A.* **65**, 19209.
112. Meleshko, K. V. (1959). *Vop. Pitan.* **18**, 57–61; (1959). *C.A.* **53**, 18319.
113. Correa, R. (1957). *Arqs Inst. Biol.* **24**, 199–227; (1960). *C.A.* **54**, 21359.
114. Mozgovaya, E. N. and Arnautov, N. V. (1960). *Izv. Sib. Otdel. Akad. Nauk S.S.S.R.* No. 2, 104–10; (1961). *C.A.* **55**, 1845.
115. Lieldiens, R., Dzenite, A., and Liberts, V. (1969). *Latv. Lopkopibas Vet. Zināt. Petnieciska Inst. Raksti* **22**, 59–67; (1970). *C.A.* **73**, 11853.
116. Shen, H. I. (1961). *Trudȳ Mosk. Vet. Akad.* **36**, 118–26; (1964). *C.A.* **60**, 7222.
117. Avakyan, A. S. (1968). *Izv. Sel'.-khoz. Nauk, Min. Sel'. Khoz. Arm. S.S.R.* **11**, 119–23; (1968). *C.A.* **69**, 75015.
118. Kondrat'ev, Yu. N. (1969). *Trudȳ Smolensk. Nauch.-Issled. Vet. Sta.* No. 3, 74–9; (1973). *C.A.* **78**, 69601.

119. Kneta, Z. (1963). *Latv. PSR Zināt. Akad. Vest.* No. 9, 115–22; (1964). *C.A.* **60**, 7221.
120. Kirchgessner, M. (1959). *Z. Tierphysiol. Tierernähr. Futtermittelk.* **14**, 214–17; (1960). *C.A.* **59**, 16577.
121. Eyubov, I. Z. (1967). *Veterinariya* **44**, 98–100; (1968). *C.A.* **68**, 1220.
122. Rish, M. A., Ben-Utyaeva, G. S., and Shimanov,V. G.(1958). *Nauch. Tr. Nauch.- Issled. Inst. Karakulevod. Uzbek. Akad. Sel'.-khoz. Nauk* **7**, 249–61 ; (1959). *C.A.* **53**, 22361.
123. Dymko, E. F. (1966). *Vest. Sel'.-khoz. Nauki* **9**, 47–51 ; (1967). *C.A.* **66**, 819.
124. Pavkovi-Filipovi, Z. (1961). *Archo Farmacog.* **11**, 7–9; (1962). *C.A.* **56**, 12057.
125. Mokranjac, M. S. and Soldatovic, D. (1963). *Acta Pharm. Jugoslav.* **13**, 43–50; (1963). *C.A.* **59**, 13253.
126. Gili, G., Toscano, G., and Lai, P. (1959). *Ann. Fac. Med. Vet.*, *Torino* **9**, 281–9; (1961). *C.A.* **55**, 11586.
127. Werner, A. and Anke, M. (1960). *Arch. Tierernähr.* **10**, 42–53; (1960). *C.A.* **54**, 16574.
128. Taucins, E. and Svilane, A. (1965). *Trudȳ Lab. Biokhim. Fiziol. Zhivotn., Inst. Biol. Akad. Nauk Latv. S.S.R.* **4**, 247–50; (1967). *C.A.* **66**, 35784.
129. Kiermeier, F. and Winkelman, H. (1961). *Z. Lebensmitterlunters. u.-Forsch.* **115**, 309–22; (1962). *C.A.* **56**, 742.
130. Leonov, V. A., Terent'eva, M. V., and Gorski, N. A. (1960). *Vest. Akad. Navuk Belaruss. S.S.R., Ser. Biyal. Navuk* No. 3, 47–55; (1962). *C.A.* **56**, 15843.
131. Odynets, R. N. and Valuiskii, P. P. (1959). *Izv. Akad. Nauk Kirg. S.S.R.* **1**, *Ser. Biol. Nauk* No. 1, 127–38; (1960). *C.A.* **54**, 15569.
132. Vsyakikh, M. I. (1959). Int. Dairy Cong., Proc. 15th Cong., London, Vol. 3, 1761–5; (1960). *C.A.* **54**, 12411.
133. Shutov, I. S. (1957). *Trudȳ Ul'yanovsk. Sel'.-khoz. Inst.* **5**, 143–56; (1961). *C.A.* **55**, 7571.
134. Berzins, J. (1958). *Latv. PSR Zināt. Akad. Vest.* No. 7, 35–8; (1959). *C.A.* **53**, 9399.
135. Panova, S. V. (1962). *Nauch. Tr. Nauch.-Issled. Inst. Zhivotnovod. Lesostepi Poles'ya* **32**, 45–53; (1963). *C.A.* **59**, 11957.
136. Grozhevskaya, S. B. (1964). *Trudȳ Permsk. Sel'.-khoz. Inst.* **21**, 13–29; (1966). *C.A.* **64**, 5529.
137. Matochkin, M. I. (1966). *Trudȳ Troitsk. Vet. Inst.* **11**, 45–50; (1967). *C.A.* **67**, 80144.
138. Poppe, S. (1966). *Wiss. Z. Univ. Rostock, Math.-Naturwiss. Reihe* **15**, 825–9; (1967). *C.A.* **66**, 113542.
139. Balla, I. (1967). *Mosonmagy. Agrartud. Foiskola Kozlem.* **10**, 207–12; (1969). *C.A.* **71**, 1183.
140. Simbinov, Sh. D. and Kusebaeva, F. Ya. (1968). Mater. Konf. Fiziol. Respub. Srednei Azii Kaz. 4th, Vol. 2, 223–5; (1973). *C.A.* **78**, 3041.
141. Shatokhina, A. P. (1968). *Dokl. TSKHA* No. 135, 79–83; (1969). *C.A.* **70**, 55307.
142. Vaskov, B., Petkov, K., and Pesevska, V. (1971). *Jugoslav. Physiol. Pharmacol. Acta* **7**, 301–5; (1972). *C.A.* **76**, 98277.
143. Gugla, V. G. and Skukovskii, B. A. (1972). Fiziol. Osn. Povysh. Prod. Zhivotn., 19–24; (1975). *C.A.* **82**, 123683.
144. Simbinov, Sh. D. (1972). *Trudȳ Inst. Fiziol., Akad Nauk Kazak. S.S.R.* **17**, 90–3; (1973). *C.A.* **78**, 83092.
145. Oll, U. (1961–2). *Jahr. Arbeitsgemeinsch Fuetterungsberat.* **4**, 79–96; (1965). *C.A.* **62**, 858.

146. Tokovoi, N. and Lapshina, L. (1962). *Trudy Krasnoyarsk. Sel'.-khoz. Inst.* **13**, 69–79; (1964). *C.A.* **60**, 14923.
147. Korshakov, P. N. (1958). *Zap. Voronezh. Sel'.-khoz. Inst.* **28**, 210–12; (1959). *C.A.* **53**, 22608.
148. Kondratenko, F. M. (1974). *Visn. Sil's'hogospod. Nauki* No. 11, 94–100; (1975). *C.A.* **82**, 138078.
149. Buevich, E. M. (1962). *Trudy Troitsk. Vet. Inst.* **8**, 50–8; (1963). *C.A.* **58**, 938.
150. Pimenov, P. K. (1968). *Trudy Ul'yanovsk. Sel'.-khoz. Inst.* **13**, 317–24; (1971). *C.A.* **74**, 29507.
151. Palo, V. (1962). *Sb. Prác Chem. Fak. SVST* No. 2, 107–15; (1963). *C.A.* **59**, 9243.
152. Simbinov, Sh. D. (1969). *Izv. Akad. Nauk Kazak. S.S.R., Ser. Biol.* **7**, 56–9; (1970). *C.A.* **72**, 108465.
153. Avakyan, A. S. (1975). *Trudy Stavrop. Sel'.khoz. Inst.* **38**, 88–90; (1976). *C.A.* **85**, 175876.
154. Boganov, G. A. (1958). *Zhivotnovodstvo* No. 1, 42–6; (1961). *C.A.* **55**, 26154.
155. Kholmanov, M. A., Shutov, I. S., Kakurina, A. G., Al'binskaya, N. N., Medvedeva, N. K., and Romenskaya, T. F. (1959). *Trudy Ul'yanovsk. Sel'.-khoz. Inst.* **7**, 23–9; (1961). *C.A.* **55**, 12574.
156. Lapshin, S. A. (1969). *Uchen. Zap. Mord. Gos. Univ.* No. 77, 3–8; (1971). *C.A.* **74**, 108827.
157. Moiseev, V. A. (1964). *Izv. Akad. Nauk S.S.R., Ser. Biol. Nauk* No. 1, 89–92; (1965). *C.A.* **62**, 12250.
158. Moiseev, V. A. and Sidorchuk, R. T. (1968). *Trudy Inst. Fiziol. Akad. Nauk Kazak. S.S.R.* **12**, 120–6; (1970). *C.A.* **73**, 32876.
159. Grachev, A. D. and Rodina, A. M. (1969). Mikroelem. Biosfere Primen Ikh. Sel. Khoz. Med. Sib. Dal'nego Vostoka, Dokl. Sib. Konf. 3rd, 358–61; (1973). *C.A.* **79**, 114257.
160. Arandi, P. and Luht, A. (1962). *Loodusuur. Seltsi Aastar.* **55**, 263–75; (1964). *C.A.* **60**, 14897.
161. Nakajima, T. and Moriwaki, K. (1971). *Shiga Kenritsu Tanki Daigaku Gakujutsu Zasshi* No. 12, 38–42; (1972). *C.A.* **76**, 32596.
162. Andrews, E. D. (1958). *New Zealand J. Agric.* **97**, 429–30.
163. Skerman, K. D., Sutherland, A. K., O'Halloran, M. W., Bourke, J. M., and Munday, B. L. (1959). *Am. J. Vet. Res.* **20**, 977–84.
164. Skerman, K. D., O'Halloran, M. W., and Munday, B. L. (1961). *Aust. Vet. J.* **37**, 181–4.
165. O'Halloran, M. W. and Skerman, K. D. (1961). *Brit. J. Nutr.* **15**, 99–108.
166. Berzins, J. (1967). *Latv. PSR Zināt. Akad. Vestis* No. 7, 142–3; (1967). *C.A.* **67**, 19043.
167. Howard, D. A. (1970). *Vet. Record* **87**, 771–4.
168. Baiturin, M. A., Rakishev, N., and Slamkulov, O. (1972). *Trudy Alma-Atinsk. Zoovet. Inst.* **24**, 94–7; (1974). *C.A.* **81**, 118868.
169. Clanton, D. C. and Rowden, W. W. (1963). *J. Range Mgmt* **16**, 16–17.
170. Alexander, G. I., Harvey, J. M., Lee, J. H., and Stubbs, W. C. (1967). *Aust. J. Agric. Res.* **18**, 169–81.
171. Andrews, E. D., Hart, L. I., and Stephenson, B. J. (1958). *Nature, Lond.* **182**, 869–70.
172. Andrews, E. D., Hart, L. I., and Stephenson, B. J. (1959). *New Zealand J. Agric. Res.* **2**, 274–82.
173. Andrews, E. D. and Hart, L. I. (1962). *New Zealand J. Agric. Res.* **5**, 403–8.

174. Andrews, E. D., Hart, L. I., and Stephenson, B. J. (1960). *New Zealand J. Agric. Res.* **3**, 364–76.
175. Somers, M. and Gawthorne, J. M. (1969). *Aust. J. Exp. Biol. Med. Sci.* **47**, 227–33.
176. Koval'skii, V. V. and Raetskaya, Yu. I. (1960). *Trudy Biogeokhim. Lab., Akad. Nauk S.S.S.R.* **11**, 102–8; (1961). *C.A.* **55**, 21284.
177. Zherebtsov, P. I., Vrakin, V. F., and Khodyrev, A. A. (1970). *Dokl. TSKHA* No. 157, 241–6; (1971). *C.A.* **74**, 73496.
178. Raetskaya, Yu. I. (1962). Mikroelementy v Sel'. Khoz. i Med. Ukr. Nauch-.-Issled. Inst. Fiziol. Rast., Akad. Nauk Ukr. S.S.R., Materialy 4-go (Chetvertogo) Vses. Soveshch., Kiev, 511–15; (1965). *C.A.* **63**, 8796.
179. Polukhin, F. S. and Savoiskic, A. G. (1962). *Trudy Mosk. Vet. Akad.* **44**, 116–23; (1965). *C.A.* **63**, 8797.
180. Svegzdaite-Laurinaviciene, D. (1964). *Liet. TSR Mokslic Akad. Darb., Ser. C.* No. 3, 3–9; (1965). *C.A.* **62**, 9571.
181. Hedrich, M. F., Elliot, J. M., and Lowe, J. E. (1973). *J. Nutr.* **103**, 1646–51.
182. Modyanov, A. V., Raetskaya, Yu. I., and Kholmanov, A. M. (1962). *Trudy Vses. Nauch.-Issled. Inst. Zhivotnovod.* **24**, 203–13; (1963). *C.A.* **59**, 13154.
183. Inoue, R. (1971). *Kobe Daigaku Nogakubu Kenkyu Hokoku* **9**, 111–22; (1972). *C.A.* **76**, 33056.
184. Inoue, R. (1971). *Kobe Daigaku Nogakubu Kenkyu Hokoku* **10**, 133–42; (1973). *C.A.* **78**, 83082.
185. Uesaka, S., Kawashima, R., and Zembayashi, M. (1965). *Kyoto Daigaku Shokuryo Kagaku Kenkyusho Hokoku* No. 28, 10–18; (1965). *C.A.* **63**, 4759.
186. Odynets, R. N. and Minchuk, N. P. (1969). *Mikroelem. Zhivotnovod. Rastenievod.* No. 8, 3–14; (1970). *C.A.* **73**, 106832.
187. Gawthorne, J. M. (1970). *Aust. J. Exp. Biol. Biol. Med. Sci.* **48**, 285–92.
188. Gawthorne, J. M. (1970). *Aust. J. Exp. Biol. Med. Sci.* **48**, 293–300.
189. Dreyfus, J. C. (1967). *Traite Biochem. Gen.* **3**, 574–7; (1967). *C.A.* **67**, 51592.
190. MacPherson, A. and Moon, F. E. (1973). Trace Element Metab. Anim., Proc. Int. Symp. 2nd, 624–7; (1975). *C.A.* **82**, 56328.
191. Lapshina, L. N. (1963). *Trudy Krasnoyarsk. Sel'.-khoz. Inst.* **16**, 96–103; (1965). *C.A.* **62**, 15145.
192. Grozhevskaya, S. B. (1963). *Trudy Permsk. Gos. Sel'.-khoz. Inst.* **18**, 57–65; (1965). *C.A.* **63**, 6075.
193. Gashimov, A. A. and Dadashev, Ch. N. (1967). Mater. Nauch. Konf. Azerb. Nauch.-Issled. Inst. Zhivotnovod., 56–8; (1969). *C.A.* **71**, 121037.
194. Zamanov, I. B. and Nagiev, I. M. (1968). Mater. Respub. Konf. Probl. "Mikroelem. Med. Zhivotnovod." 1st, 123–4; (1971). *C.A.* **74**, 29472.
195. Nefedov, N. A. and Aksenova, T. E. (1968). *Uchen. Zap. Kazansk. Vet. Inst.* **99**, 210–17; (1970). *C.A.* **72**, 129705.
196. Vas'kovich, Ya. V. (1968). *Sb. Stud. Nauch. Rab., Mosk. Sel'.-khoz. Akad.* No. 16, 342–5; (1970). *C.A.* **72**, 10412.
197. Tarasova, A. N. (1962). Mikroelementy v Zhivotnovod. Min. Sel'.-khoz. Khoz. S.S.S.R., Vses. Akad. Sel'.-khoz. Nauk, Otdel. Zhivotnovod., 53–61; (1963). *C.A.* **58**, 11738.
198. Rys, R. and Krelowska, M. (1963). *Zesz. Probl. Postep. Nauk Roln.* No. 41, 127–31; (1964). *C.A.* **60**, 16276.
199. Gadzhiev, F. M. (1966). *Izv. Akad. Nauk Azerb. S.S.R., Ser. Biol. Nauk* No. 4, 90–7; (1967). *C.A.* **66**, 63031.

200. Gadzhiev, F. M. (1968). Mater. Respub. Konf. Probl. "Mikroelem. Med. Zhivotnovod." 1st, 110–12; (1971). *C.A.* **74**, 29470.
201. Koval'skii, V. V., Karaev, A. I., and Gadzhiev, F. M. (1968). Mater. Respub. Konf. Probl. "Mikroelem. Med. Zhivotnovod." 1st, 129–30; (1971). *C.A.* **74**, 29480.
202. Gadzhiev, F. M. and Ifraimova, Z. N. (1968). Mater. Respub. Konf. Probl. "Mikroelem. Med Zhivotnovod." 1st, 113–14; (1971). *C.A.* **74**, 29472.
203. Nikitin, B. N., Il'ina, K. A., and Moiseev, V. A. (1968). *Trudy Inst. Fiziol., Akad. Nauk Kazak. S.S.R.* **12**, 136–43; (1970). *C.A.* **73**, 32877.
204. Odynets, R. N., Tokobaev, E. M., Aituganov, M. D., and Dedova, R. G. (1971). *Mikroelem. Zhivotnovod. Rastenievod.* No. 10, 26–42; (1973). *C.A.* **79**, 90736.
205. Pyak, E. S. (1972). *Vest. Sel'.-khoz. Nauki* **15**, 98–100; (1973). *C.A.* **78**, 3044.
206. Inoue, R. and Onishi, M. (1964). *Hyogo Noka Daigaku Kenkyu Hokoku, Chickusan-gaku Hen* **6**, 89–94; (1966). *C.A.* **64**, 8706.
207. Holmes, E. G. (1965). *J. Exp. Physiol.* **50**, 203–13.
208. Koval'skii, V. V. and Rambidi, M. I. (1958). *Dokl. Vses. Akad. Sel'.-khoz. Nauk im V.I. Lenina* **23**, 29–33; (1959). *C.A.* **53**, 7339.
209. Rambidi, M. I. (1962). *Trudy Vses. Nauch.-Issled. Inst. Zhivotnovod.* **24**, 258–68; (1963). *C.A.* **59**, 13154.
210. Ayupova, R. S. (1966). *Trudy Inst. Fiziol. Akad. Nauk Kazak. S.S.R.* **10**, 114–18; (1967). *C.A.* **67**, 19050.
211. Kabysh, A. A. (1962). *Trudy Troitsk. Vet. Inst.* **8**, 65–70; (1963). *C.A.* **59**, 10526.
212. Valyushkin, K. D. (1974). *Vest. Akad. Navuk Belaruss. S.S.R., Ser. Sel'skagaspad. Navuk* No. 1, 94–6; (1974). *C.A.* **81**, 2592.
213. Rozybakiev, M. A. (1968). Mater. Respub. Konf. Probl. "Mikroelem. Med. Zhivotnovod." 1st, 155–7; (1971). *C.A.* **74**, 29491.
214. Rozybakiev, M. A. (1968). Mater. Konf. Fiziol. Respub. Srednei Azii Kazak. 4th, Vol. 2, 299–3; (1973). *C.A.* **78**, 96399.
215. Rozybakiev, M. A., Ababkov, M. M., and Dzhubanova, G. R. (1968). *Trudy Inst. Fiziol., Akad. Nauk Kazak. S.S.R.* **12**, 39–45 (1970). *C.A.* **73**, 42752.
216. Rozybakiev, M. A. (1964). *Trudy Inst. Fiziol. Akad. Nauk Kazak. S.S.R.* **7**, 88–96; (1966). *C.A.* **64**, 5528.
217. Ababkov, M. M. (1968). Mater. Respub. Konf. Probl. "Mikroelem. Med. Zhivotnovod." 1st, 95–6; (1971). *C.A.* **74**, 29462.
218. Berzins, J. (1955). Mikroelementy v Sel'. Khoz. i Med., Akad. Nauk Latv. S.S.R., Otdel. Biol. Nauk, Tr. Vses. Soveshchan. Riga, 511–27; (1959). *C.A.* **53**, 11560.
219. Muganlinskaya, D. I. (1968). Mater. Respub. Konf. Probl. "Mikroelem. Med. Zhivotnovod." 1st, 145–7; (1971). *C.A.* **74**, 29486.
220. Zaderii, I. I., Knyazeva, A. P., Nesterenko, K. E., Anisimov, E. V., and Korzhukhovskaya, S. Yu. (1965). Mikroelementy v Zhivotnovod. i Med., Akad. Nauk Ukr. S.S.R., Resp. Mezhvedomstv. Sb., 36–43; (1966). *C.A.* **64**, 14658.
221. Andrews, E. D., Hogan, K. G., Stephenson, B. J., White, D. A., and Elliott, D. C. (1970). *New Zealand J. Agric. Res.* **13**, 950–65.
222. Aituganov, M. D. (1969). *Mikroelem. Zhivotnovod. Rastenievod.* No. 8, 24–5; (1970). *C.A.* **73**, 106835.
223. Aituganov, M. D. (1968). Mater. Respub. Konf. Probl. "Mikroelem. Med. Zhivotnovod." 1st, 105; (1971). *C.A.* **74**, 29808.
224. Koval'skii, V. V. and Pen'kova, A. F. (1968). *Dokl. Vses. Akad. Sel'.-khoz. Nauk* **9**, 27–9; (1969). *C.A.* **70**, 45237.

225. Kusen, S. I., Porodko, I. S., and Dorda, V. Ya. (1964). *Dopov. Akad. Nauk Ukr. S.S.R.* No. 7, 950–2; (1964). *C.A.* **61**, 11077.
226. Zaderiff, I. I. (1955). Mikroelementy v Sel'. Khoz. i Med., Akad. Nauk Latv. S.S.R., Otdel. Biol. Nauk, Tr. Vses. Soveshchan. Riga, 611–15; (1959). *C.A.* **53**, 11561.
227. Andrews, E. D., Grant, A. B., and Stephenson, B. J. (1964). *New Zealand J. Agric. Res.* **7**, 17–27.
228. Grace, N. D. (1975). *Brit. J. Nutr.* **34**, 73–82.
229. Berzins, J. and Rozenbachs, J. (1953). *Latv. PSR Zināt. Akad. Vest.* No. 9, 39–46; (1954). *C.A.* **48**, 12260.
230. Valdamis, A., Taucins, E., Svilane, A., and Grundman, G. A. (1965). *Trudȳ Lab. Biokhim. Fiziol. Zhivotn., Inst. Biol. Akad. Nauk Latv. S.S.R.* **4**, 149–61; (1966). *C.A.* **65**, 19061.
231. Andreevskaya, V. E. (1967). *Probl. Ekol.* **1**, 223–6; (1970). *C.A.* **72**, 40298.
232. Baiturin, M. A. and Tanatarov, A. B. (1968). *Vest. Sel'.-khoz. Nauki* **11**, 73–6; (1968). *C.A.* **69**, 25368.
233. Fezliev, Z. G. and Kokov, T. N. (1969). *Sb. Nauch. Rab. Dagestan. Nauch.-Issled. Vet. Inst.* **3**, 214–21; (1971). *C.A.* **75**, 96009.
234. Taucins, E., Svilane, A., and Valdamis, A. (1969). Fiziol. Aktiv. Komponenty Pitan. Zhivotn., 185–97; (1971). *C.A.* **74**, 29458.
235. Konopatov, Yu. V. and Gurevich, D. I. (1972). *Mater. Vses. Nauch. Soveshch. Konf. Vses. Nauch.-Issled. Tekhnol. Inst. Ptitsevod.* No. 5, 235–9; (1974). *C.A.* **80**, 107011.
236. Kanimetov, A. K. and Fateev, V. A. (1973). *Mikroelem. Zhivotnovod. Rastenievod.* **12**, 54–66; (1975). *C.A.* **83**, 57065.
237. Fateev, V. A. (1973). *Sb. Tr. Aspir. Molodykh Uch.-Kirg. Nauch.-Issled. Inst. Zhivotnovod. Vet.* **5**, 69–71; (1974). *C.A.* **83**, 130345.
238. Shkunkova, Yu. S., Thachuk, V. G., and Pyshnik, V. F. (1974). *Trudȳ Beloruss. Nauch.-Issled. Inst. Zhivotnovod.* **14**, 194–202; (1975). *C.A.* **83**, 130329.
239. Ko, Y. D., Song, D. J., and Ha, J. K. (1975). *Hanguk Ch'uksan Hakhoe Chi* **17**, 90–3; (1976). *C.A.* **84**, 149636.
240. Polyakova, E. P. (1972). *Dokl. TSKHA* No. 185, 139–42; (1973). *C.A.* **78**, 146540.
241. Tastaldi, H., Melardi, E. B., Leal, A., and Buccheri, A. (1954). *Anais Fac. Farm. Odont. Univ. Sao Paulo* **12**, 172–82; (1956). *C.A.* **50**, 9535.
242. Burns, M. J. and Salmon, W. D. (1956). *J. Agric. Food Chem.* **4**, 257–9.
243. Panic, B., Stosic, D., Hristic, B., and Cuperlovic, M. (1964). Radioisotopes Animals Nutr. Physiol. Proc. Symp., Prague, 387–98; (1966). *C.A.* **64**, 16357.
244. Menke, K. H. (1959). *Arch. Geflügelk.* **23**, 32–8; (1961). *C.A.* **55**, 14609.
245. Kawecki, A. and Rotenberg, S. (1963). *Acta Physiol. Polon.* **14**, 441–53; (1965). *C.A.* **63**, 15271.
246. Veliky, I. (1961). *Vet. Casopis* **10**, 450–6; (1962). *C.A.* **56**, 3879.
247. Gadzhiev, F. M. and Alikishibekova, Z. M. (1960). *Trudȳ Sektora Fiziol., Akad. Nauk Azerb. S.S.R.* **3**, 52–5; (1962). *C.A.* **56**, 1854.
248. Gadzhiev, F. M. and Alikishibekova, Z. M. (1960). *Trudȳ Sektora Fiziol., Akad. Nauk Azerb. S.S.R.* **3**, 93–6; (1962). *C.A.* **56**, 1855.
249. Rybina, E. V. and Uzilevskaya, P. S. (1962). *Trudȳ Nauch.-Issled. Inst. Zhivotnovod. Uzbek. Akad. Sel'.-khoz. Nauk* No. 7, 111–17; (1963). *C.A.* **59**, 4318.
250. Menke, K. H. (1964). *Z. Physiol. Chemie* **336**, 257–63; (1964). *C.A.* **61**, 15097.
251. Menke, K. H. and Marquering, B. (1964). Radioisotopes Animal Nutr. Physiol. Proc. Symp., Prague, 373–86; (1966). *C.A.* **64**, 16357.

252. Apsite, M. (1968). Mikroelem. Organizme Ryb Ptits, 63–83; (1970). *C.A.* **72**, 40292.
253. Apsite, M. and Bermane, S. (1969). *Uchen. Zap. Latv. Gos. Univ.* **100**, 36–60; (1971). *C.A.* **75**, 46150.
254. Konopatov, Yu. V. (1971). *Sb. Rab. Leningrad. Vet. Inst.* No. 32, 363–7; (1973). *C.A.* **79**, 30733.
255. Mel'chenko, A. I. (1972). *Dokl. TSKHA* No. 185, 143–6; (1973). *C.A.* **78**, 146535.
256. Konopatov, Yu. V. (1974). *Sb. Rab. Leningrad. Vet. Inst.* **36**, 53–7; (1975). *C.A.* **82**, 71867.
257. Bolotnikov, I. A. and Konopatov, Yu. V. (1973). *Sb. Rab. Leningrad. Vet. Inst.* **33**, 20–4; (1974). *C.A.* **81**, 168097.
258. Maturova, E. T. (1964). Mikroelem. Biosfere Ikh Primen. Sel'. Khoz. Med. Sib. Dal'nego Vostoka, Dokl. Sib. Konf. 2nd, 501–2; (1969). *C.A.* **70**, 85390.
259. Kirkpatrick, D. C. and Coffin, D. E. (1975). *J. Sci. Food Agric.* **26**, 99–103.
260. Khamidullina, L. (1974). *Vop. Pitan.* No. 4, 79–80; (1975). *C.A.* **83**, 162318.
261. Young, R. S. (1961). *J. Agric. Food Chem.* **8**, No. 6, 485–6.
262. Berzins, J. (1951). *Latv. PSR Zināt. Akad. Vest.* 415–20; (1953). *C.A.* **47**, 12440.
263. Dinusson, W. E., Klosterman, E. W., Lasley, E. L., and Buchanan, M. L. (1953). *J. Anim. Sci.* **12**, 623–7.
264. Viana, J. A. C. and Moreira, H. A. (1956). *Arqos Esc. Super. Vet. Est. Minas Gerais* **9**, 161–8; (1958). *C.A.* **52**, 5573.
265. Shergin, N. P. (1957). *Trudȳ Vses. Nauch.-Issled. Inst. Zhivotnovod.* **21**, 229–39; (1959). *C.A.* **53**, 15246.
266. Rechka, J. and Kalous, J. (1960). *Sb. Vys. Sk. Zeměd. Brne, Rada A*, 263–75; (1961). *C.A.* **55**, 16711.
267. Kalous, J. and Loudil, L. (1960). *Sb. Českoslov. Akad. Zeměd. Ved, Zivocisna Vyroba* **5**, 11–20; (1961). *C.A.* **55**, 2827.
268. Admina, L. Ya. and Panova, S. V. (1966). *Sel'.-khoz. Biol.* **1**, 766–9; (1967). *C.A.* **66**, 63041.
269. Vovsova, R. (1976). *Biol. Chem. Vyz. Zuirat* **12**, 53–9; 61–6; (1976). *C.A.* **85**, 31849–50.
270. Rastegaev, Yu. M. (1976). *Svinovodstvo* No. 5, 18; (1976). *C.A.* **85**, 76567.
271. Petrukhin, I. V. (1967). *Vest. Sel'.-khoz. Nauki* **12**, 86–9; (1967). *C.A.* **67**, 80146.
272. Loudil, L. and Kalous, J. (1960). *Sb. Českoslov. Akad. Zeměd. Ved, Zivocisna vyroba* **5**, 663–70; (1961). *C.A.* **55**, 4680.
273. Los'makova, S. I. (1971). Pishchevarenie Obmen Veskchestv Svinei, 196–200; (1972). *C.A.* **77**, 60390.
274. Tmenov, I. D. (1964). *Trudȳ Sev.-Osetinsk. Sel'.-khoz. Inst.* **22**, 131–7; (1965). *C.A.* **63**, 8797.
275. Shuopaitite, G. (1957). *Byull. Nauch.-Tekh. Inform. Litovsk. Nauch.-Issled. Inst. Zhivotnovod. Vet.* No. 2, 17–18; (1959). *C.A.* **53**, 1482.
276. Kozhemyakin, N. G. and Shestakov, Yu. M. (1975). *Sb. Rab. Leningrad. Vet. Inst.* **42**, 109–12; (1976). *C.A.* **84**, 149635.
277. Grzeszczak-Swietlikowska, U. (1961). *Zesz. Probl. Postep. Nauk Roln.* No. 28, 139–47; (1964). *C.A.* **61**, 16512.
278. Bularga, I. A. (1973). *Trudȳ Kishinev. Sel'.-khoz. Inst.* **113**, 85–8; (1975). *C.A.* **82**, 29975.
279. Kunev, M. and Pazardzhiev, A. (1975). *Zhivotnovod. Nauki* **12**, 61–5; (1975). *C.A.* **83**, 41852.
280. Krylova, V. S. (1959). *Vest. Sel'.-khoz. Nauki, Vses. Akad. Sel'.-khoz. Nauk* No. 9, 80–7; (1962). *C.A.* **54**, 10076.

281. Korshun, A. I. (1962). *Trudȳ Troitsk. Vet. Inst.* **8**, 74–7; (1964). *C.A.* **60**, 11122.
282. Raetskaya, Yu. I. and Netecha, V. I. (1968). Mater. Respub. Konf. Probl. "Mikroelem. Med. Zhivotnovod." 1st, 149–51; (1971). *C.A.* **74**, 29487.
283. Raetskaya, Yu. I., Agapitova, G. N., Veselitskaya, T. V., and Netecha, V. I. (1972). *Trudȳ Vses. Nauch.-Issled. Inst. Zhivotnovod.* No. 34, 290–7; (1974). *C.A.* **80**, 46677.
284. Stepurin, G. F. and Svezhentsov, A. I. (1969). *Trudȳ Kishinev. Sel'.-khoz. Inst.* **58**, 72–81; (1971). *C.A.* **74**, 108830.
285. Burch, R. E., Williams, R. V., and Sullivan, J. F. (1973). *Am. J. Clin. Nutr.* **26**, 403–8.
286. Khokhrin, S. N. (1968). *Kormlenie Sel'.-khoz. Zhivotn.* No. 8, 193–201; (1970). *C.A.* **72**, 52410.
287. Kneta, Z. Ya., Zizum, A. I., and Shmidt, A. A. (1965). Biokhim. Faktory Reaktivnost Organizma, Latv. Inst. Eksp. Klin. Med. Akad. Med. Nauk S.S.S.R., 93–112; (1967). *C.A.* **67**, 62054.
288. Genci, L. and Galik, J. (1964). *Pol'nohospodarstvo* **10**, 633–6; (1965). *C.A.* **62**, 4389.
289. Taucins, E. and Svilane, A. (1965). *Fiziol. Zhivotn. Inst. Biol., Akad. Nauk Latv. S.S.R.* **4**, 251–4; (1967). *C.A.* **66**, 17385.
290. Pushkarev, R. P. (1973). Mater. Povolzh. Konf. Fiziol. Uchastiem Biokhim. Farmakol. Morfol. 6th, Vol. 1, 300–1; (1975). *C.A.* **82**, 123670.
291. Ierusalimskii, I. G. (1958). *Priroda* **47**, 104–5; (1959). *C.A.* **53**, 4452.
292. Kichina, M. M. (1960). *Trudȳ Mosk. Vet. Akad.* **30**, 169–70; (1962). *C.A.* **56**, 2753.
293. Kichina, M. M. (1961). *Vest. Akad. Navuk Belaruss. S.S.R., Ser. Biyal. Navuk* No. 1, 98–104; (1965). *C.A.* **62**, 5630.
294. Skoropostizhnaya, A. S. (1958). *Fiziol. Zh., Akad. Nauk Ukr. S.S.R.* **4**, 537–41; (1959). *C.A.* **53**, 2389.
295. Dymko, E. F. (1968). *Trudȳ Alma-Atinsk. Zootek.-Vet. Inst.* **15**, 39–44; (1971). *C.A.* **74**, 29938.
296. Skoropostizhnaya, A. S. (1962). *Vrach. Delo* No. 1, 165–7; (1962). *C.A.* **56**, 15910.
297. Sinnett, K. I. and Spray, G. H. (1965). *Brit. J. Nutr.* **19**, 119–23.
298. Savrich, V. O. (1961). *Ukr. Biokhim. Zh.* **33**, 732–7; (1962). *C.A.* **56**, 6442.
299. Krantz, S., Bartolomaeus, A., and Lober, M. (1974). *Folia Haematol.* **101**, 785–91; (1975). *C.A.* **82**, 41163.
300. Yastrebov, A. P. (1965). *Patol. Fiziol. Eksp. Terapi.* **9**, 34–7; (1967). *C.A.* **66**, 17381.
301. Pavero, A. and Garello, L. (1958). *Arch. "E. Maragliano" Patol. Clin.* **14**, 1081–9; (1959). *C.A.* **53**, 6399.
302. Babadzhanov, S. N. (1973). *Uzbek. Biol. Zh.* **17**, 71–2; (1974). *C.A.* **80**, 118571.
303. Kichina, M. M. (1966). *Nauch. Dokl. Vȳssh. Shk., Biol. Nauki* No. 1, 74–7; (1966). *C.A.* **65**, 2887.
304. Boechko, F. F. (1970). *Vop. Med. Khim.* **16**, 607–12; (1971). *C.A.* **74**, 51087.
305. Kichina, M. M. (1970). *Uchen. Zap. Vitebsk. Vet. Inst.* **22**, 153–6; (1973). *C.A.* **78**, 28254.
306. Kichina, M. M. (1969). *Uchen. Zap. Vitebsk. Vet. Inst.* **21**, 45–9; (1970). *C.A.* **72**, 40302.
307. Krantz, S. and Lober, M. (1973). *Folia Haematol.* **100**, 295–302; (1974). *C.A.* **80**, 94418.

308. Davydov, N. I. and Reshetovskaya, N. A. (1963). *Sb. Nauch. Rab. Ryazansk. Sel'.-khoz. Inst.* No. 1, 140–5; (1965). *C.A.* **62**, 5792.

309. Kornienko, V. V. (1969). *Oftal. Zh.* **24**, 618–20; (1970). *C.A.* **73**, 96294.

310. Sasmal, N., Mukherjee, D., Kar, N. C., and Chatterjee, G. C. (1968). *Indian J. Biochem.* **5**, 123–5.

311. Hasegawa, T. (1974). *Nippon Eiseigaku Zasshi* **29**, 289–99; (1975). *C.A.* **82**, 165463.

312. Gorshkov, S. I. and Denisov, A. F. (1962). Reaktsii Organizma na Deistve Malykh Doz Ioniziruyshchei Radiatsii, Akad. Nauk S.S.S.R., 226–32; (1963). *C.A.* **58**, 14539.

313. Fedosova, E. E. (1968). *Mikroelement. Med.* No. 1, 181–3; (1969). *C.A.* **71**, 47332.

314. Dufrenoy, J. (1959). *Rev. Pathol. Gen. Physiol. Clin.* **59**, 451–2; (1959). *C.A.* **53**, 19075.

315. Abrarov, A. A. (1961). *Vop. Pitan.* **20**, 39–44; (1962). *C.A.* **57**, 14263.

316. Abrarov, A. A. (1962). *Med. Zh. Uzbek.* No. 3, 54–7; (1962). *C.A.* **57**, 3849.

317. Saikkonen, J. (1962). *Acta Pathol. Mikrobiol. Scand.* **55**, 129–32; (1962). *C.A.* **57**, 10143.

318. Fiedler, H. and Taube, C. (1970). *Thromb. Diath. Haemorrh.* **24**, 587–600; (1971). *C.A.* **74**, 85152.

319. Shternberg, A. I., Kusevitskii, I. A., and Abolin, E. E. (1963). *Vop. Pitan.* **22**, 41–7; (1963). *C.A.* **59**, 7932.

320. Shternberg, A. I., Kusevitskii, I. A., Aksyuk, I. N., and Abolin, E. E. (1962). *Uchen. Zap. Mosk. Nauch.-Issled. Inst. Gig.* No. 12, 57–65; (1963). *C.A.* **59**, 14349.

321. Novikova, E. P. (1963). *Vrach. Delo* No. 4, 140–2; (1963). *C.A.* **59**, 11958.

322. Jirgena, A. and Leja, D. (1966). Biokhim. Fiziol. Morfol. Obosnovaniya Diagn. Ter., Rizh. Med. Inst., 449–54; (1968). *C.A.* **68**, 57761.

323. Varma, T. N. S., Nagarajan, B., Brachmanandam, S., and Sivaramakrishnan, V. M. (1962). *Ann. Biochem. Exp. Med.* **22**, 265–78; (1963). *C.A.* **58**, 9400.

324. Schwarz, K., Roginski, E. E., and Foltz, C. M. (1959). *Nature, Lond.* **183**, 472–3.

325. Daniel, M. R., Dingle, J. T., and Lucy, J. A. (1961). *Exp. Cell Res.* **24**, 88–105.

326. Nagarajan, B., Sivaramakrishnan, V. M., and Brachmanandam, S. (1963). *Current Sci.* **32**, 8–9.

327. Brambilla, G., Boaretto, M. C., and Bolognesi, C. (1975). *Tumori* **61**, 327–32; (1976). *C.A.* **84**, 13279.

328. Levina, E. N. and Minkina, N. A. (1961). *Gig. Sanit.* **26**, 27–32; (1962). *C.A.* **56**, 3775.

329. Eichler, O., Maroske, D., Hoebel, M., and Wegener, K. (1967). *Zentbl. Biol. Aerosol-Forsch.* **13**, 535–47; (1968). *C.A.* **68**, 76628.

330. Taylor, D. M. (1962). *Phys. Med. Biol.* **6**, 445–51.

331. Carlberger, G., Magnusson, G., and Meurman, L. (1961). *Acta Med. Scand.* **170**, 479–86; (1962). *C.A.* **56**, 13389.

332. Schade, S. G., Felsher, B. F., Glader, B. E., and Conrad, M. E. (1970). *Proc. Soc. Exp. Biol. Med.* **134**, 741–3.

333. Toskes, P. P., Smith, G. W., and Conrad, M. E. (1973). *Am. J. Clin. Nutr.* **26**, 435–7.

334. Gude, Z. Zh. (1962). *Ukr. Biokhim. Zh.* **34**, 840–4; (1963). *C.A.* **58**, 11745.

335. Bryan, S. E. and Morgan, K. S. (1970). *FEBS Lett.* **9**, 277–80.

336. Lutsik, L. A. (1972). *Mikroelement. Med.* No. 3, 188–90; (1973). *C.A.* **79**, 41193.

337. Miller, A. T. and Hale, D. M. (1970). *Archs Int. Physiol. Biochim.* **78**, 475–9.
338. Kim, M. M., Khamidov, D. Kh., Nikolaev, A. I., and Pulatov, R. P. (1970). *Uchen. Tr. Gor'k. Gos. Med. Inst.* No. 32, 359–60; (1971). *C.A.* **75**, 46422.
339. Payan, H. M. (1970). *Exp. Med. Surg.* **28**, 163–8.
340. Raitses, V. S. and Pityk, N. I. (1976). *Fiziol. Zh.* **22**, 228–31; (1976). *C.A.* **84**, 133370.
341. Dingle, J. T., Heath, J. C., Webb, M., and Daniel, M. R. (1962). *Biochim. Biophys. Acta* **65**, 34–46.
342. Webb, M. (1962). *Biochim. Biophys. Acta* **65**, 47–65.
343. Bartelheimer, E. W. (1962). *Arch. Exp. Pathol. Pharmakol.* **243**, 237–53; (1962). *C.A.* **57**, 11481.
344. De Moraes, S. and Mariano, M. (1967). *Med. Pharmacol. Exp.* **16**, 441–7; (1967). *C.A.* 67, 42167.
345. Wiberg, G. S. (1968). *Can. J. Biochem.* **46**, 549–54.
346. Vasil'kov, V. V. and Golodushko, B. Z. (1971). *Dokl. Akad. Nauk Beloruss. S.S.R.* **15**, 944–6; (1972). *C.A.* **76**, 58259.
347. Khalilov, K. B. (1963). Mikroelementy v Sel'. Khoz. i Med. Sb., 534–9; (1965). *C.A.* **62**, 8178.
348. Rzakuliev, G. Ch. (1967). *Trudȳ Azerb. Nauch.-Issled. Vet. Inst.* **21**, 265–7; (1969). *C.A.* **70**, 55530.
349. Rzakuliev, G. Ch. and Mekhtiev, M. A. (1968). Mater. Respub. Konf. Probl. "Mikroelem. Med. Zhivotnovod." 1st, 153–4; (1971). *C.A.* **74**, 29489.
350. Rzakuliev, G. Ch. (1968). *Trudȳ Azerb. Nauch.-Issled. Vet. Inst.* **23**, 284–6; (1969). *C.A.* **71**, 89215.
351. Galakhova, V. N. (1959). *Nauk Zap. Stalislavs'k Med. Inst.* No. 3, 122–31; (1963). *C.A.* **59**, 10525.
352. Babenko, G. O. (1960). *Ukr. Biokhim. Zh.* **32**, 93–8; (1960). *C.A.* **54**, 21231.
353. El'tsov, N. S. (1962). Materialy 1-go (Pervogo) S'ezda Beloruss. Fiziol. Obshchestva Sb., 75–6; (1963). *C.A.* **59**, 13182.
354. Berestova, V. I., Panova, M. K., Berestov, V. A., Berestov, A. A., and Petrova, G. G. (1965). *Uchen. Zap. Petrozavodsk. Gos. Univ.* **13**, 113–17; (1966). *C.A.* **65**, 19061.
355. Berestova, V. I. and Panova, M. K. (1965). *Uchen. Zap. Petrozavodsk. Gos. Univ.* **13**, 118–22; (1966). *C.A.* **65**, 17442.
356. Bukovskaya, Z. I. (1969). *Uchen. Zap. Yakutsk. Gos. Univ.* No. 19, 36–8; (1971). *C.A.* **74**, 84764.
357. Berestova, V. I. (1969). *Uchen. Zap. Petrozavodsk. Gos. Univ.* **17**, 125–7; (1972). *C.A.* **77**, 86066.
358. Frank, C., Lamarche, M., and Kocarev, R. (1958). *J. Physiol., Paris* **50**, 284–5; (1959). *C.A.* **53**, 10424.
359. Savrich, V. O. (1961). *Ukr. Biokhim. Zh.* **33**, 266–70; (1961). *C.A.* **55**, 18913.
360. Muceniece, A. (1958). *Latv. PSR Zināt. Akad. Vest.* No. 5, 71–7; (1959). *C.A.* **53**, 5432.
361. Beskid, M. (1963). *Polska Akad. Nauk Rozprawy Wydzialu Nauk Med.* **8**, 5–36; (1964). *C.A.* **60**, 3410.
362. Beskid, M. (1967). *Folia Histochem. Cytochem.* **5**, 33–72; (1967). *C.A.* **67**, 42161.
363. Huy, N. D., Morin, P. J., Mohiuddin, S. M., and Morin, Y. (1973). *Can. J. Physiol. Pharmacol.* **51**, 46–51.
364. Mori, B. and Arikawa, T. (1965). *Nippon Nogei Kagaku Kaishi* **39**, 1–4; (1965). *C.A.* **63**, 15322.

365. Leja, D. (1965). *Latv. PSR Zināt. Akad. Vest.* No. 11, 71–5; (1966). *C.A.* **64**, 16358.
366. Jirgene, A. and Leja, D. (1967). *Latv. PSR Zināt. Akad. Vest.* No. 6, 88–93; (1967). *C.A.* **67**, 61982.
367. Kirykhin, R. A. (1958). Trudy 1-oi (Pervoi) Konf. po Molochn. Konevodstvu i Kumysodeliyu, Moscow, 175–8; (1961). *C.A.* **55**, 12690.
368. Magidov, G. A. (1962). Mikroelementy v Zhivotnovod., Min. Sel'. Khoz. S.S.S.R., Vses. Akad. Sel'.-khoz. Nauk Otdel. Zhivotnovod., 78–89; (1963). *C.A.* **58**, 11738.
369. Mikhailov, N. G. (1973). *Trudy Dal'nevost. Nauch.-Issled. Inst. Sel'. Khoz.* **13**, 412–14; (1975). *C.A.* **83**, 26620.
370. Podkarytov, F. M. (1966). *Trudy Nauch.-Issled. Inst. Sel'. Khoz. Krainego Sev.* **13**, 133–6; (1968). *C.A.* **68**, 1306.
371. Kazanovskii, E. S. (1969). *Trudy Nauch.-Issled. Inst. Sel'. Khoz. Krainego Sev.* **17**, 51–4; (1971). *C.A.* **75**, 138482.
372. Malaishkaite, B. (1955). Mikroelementy v Sel'. Khoz. i Med., Akad. Nauk Latv. S.S.R., Otdel. Biol. Nauk, Tr. Vses. Soveshch. Riga, 573–82; (1959). *C.A.* **53**, 11561.
373. Malaishkaite, B. (1957). *Byull. Nauch.-Tekh. Inform. Litovsk. Nauch.-Issled. Inst. Zhivotnovod. Vet.* No. 2, 15–16; (1959). *C.A.* **53**, 1482.
374. Mulyak, V. G. (1963). Materialy 8-oi (Vos'moi) Nauchn. Konf. po Farmakol., Moscow, Sb., 92–3; (1965). *C.A.* **63**, 8796.
375. Malaishkaite, B. (1963). *Liet. Gyvulininkystes Mokslinio Tyrimo Inst. Darb.* **6**, 209–16; (1964). *C.A.* **61**, 1026.
376. Nelson, M. M., Moyle, J. B., and Farnham, A. T. (1966). *J. Wildlife Mgmt* **30**, 423–5.
377. Baiturin, N. A. and Omarkozhaev, N. O. (1974). *Vest. Sel'.-khoz. Nauki Kaz.* **17**, 66–7; (1974). *C.A.* **81**, 118887.
378. Murakami, T., Ishihara, Y., and Uesugi, K. (1961). *Himeji Kogyo Daigaku Kenkya Hokoku* No. 13, 98–108; (1962). *C.A.* **56**, 1848.
379. Mershina, K. M., Khalina, N. M., and Krasnitskaya, A. I. (1962). Mikroelementy v Vost. Sib. i na Dal'n. Vost. Inform. Byal. Koordinats. Komis. po Mikroelementam dlya Sib. i Dal'n. Vost. No. 1, 35–43; (1963). *C.A.* **59**, 13270.
380. Chojnicka, B. and Szyszko, E. (1964). *Rocz. Państ. Zakl. Hig.* **15**, 23–6; (1964). *C.A.* **61**, 6275.
381. Katsuki, Y., Yasuda, K., Ueda, K., and Kimura, Y. (1974). *Tokyo Toritsu Eisei Kenkyusho Kenkyu Nempo* **25**, 257–63; (1975). *C.A.* **82**, 150119.
382. Tomnatik, E. N. and Batyr, A. K. (1965). *Biol. Resur. Vodoemov Mold., Akad. Nauk Mold. S.S.R. Inst. Zool.* No. 3, 64–8; (1967). *C.A.* **67**, 9138.
383. Burtov, V. Ya. (1961). *Pchelovodstvo* **38**, 22; (1964). *C.A.* **60**, 13632.
384. Goloskokov, V. G. and Kochergin, B. N. (1963). *Mikroelement. Sib., Inform. Byull.* No. 2, 81–6; (1965). *C.A.* **63**, 8797.
385. Kardakov, V. P. (1975). *Issled. Obl. Vet.* 7–9; (1976). *C.A.* **85**, 158244.
386. Chebotarev, I. I., Proskuryakova, T. R., and Lankova, T. V. (1972). *Uchen. Zap. Dal'nevost. Gos. Univ.* **57**, 80–2; (1974). *C.A.* **80**, 35927.
387. Shakhbazova, E. E. (1962). *Ukr. Biokhim. Zh.* **34**, 694–701; (1963). *C.A.* **58**, 3720.
388. Tuchkova, T. G. (1962). *Trudy Turkmen. Sel'.-khoz. Inst.* **11**, 65–70; (1965). *C.A.* **63**, 8794.

389. Kolomeitsev, N. T. and Pirogova, N. V. (1963). *Byull. Nauch.-Tekhn. Inform. Kirg. Nauch.-Issled. Inst. Zhivotnovod. Vet.* Nos. 7–8, 89–91; (1965). *C.A.* **63**, 8793.
390. Sridhara, S. and Bhat, J. V. (1966). *Proc. Indian Acad. Sci., Sect. B.* **63**, 9–16.
391. Perron, J. M., Huot, L., and Smirnoff, W. A. (1966). *Comp. Biochem. Physiol.* **18**, 869–79.
392. Becker, D. E. and Smith, S. E. (1951). *J. Animal Sci.* **10**, 266–71.
393. Andrews, E. D. (1965). *New Zealand Vet. J.* **13**, 101–13.
394. Dickson, J. and Bond, M. P. (1974). *Aust. Vet. J.* **50**, 236.
395. Wehner, A. P. and Craig, D. K. (1972). *J. Am. Ind. Hyg. Ass.* **33**, 146–55.
396. Kerfoot, E. J. (1973). U.S.N.T.I.S. PB Rep. No. 232247/7GA; (1974). *C.A.* **81**, 164229.
397. Kerfoot, E. J., Fredrick, W. G., and Domeier, E. (1975). *J. Am. Ind. Hyg. Ass.* **36**, 17–25.
398. Seidov, I. M. (1963). *Gig. Sanit.* **28**, 93–6; (1964). *C.A.* **60**, 16449.

9 Cobalt in Human Nutrition

It has been known for over fifty years that whole liver was effective in the dietary treatment of pernicious anemia. From this, and other observations, some prominent nutritionists came to the belief that minute quantities of cobalt were necessary for humans.[1-3] No indications were given, however, of the amount required.

Many years later, the pure anti-pernicious anemia factor, now known as vitamin B_{12}, was isolated, and its complete structure fully described.[4-7] Vitamin B_{12} is a red, crystalline solid, fairly soluble in water and lower alcohols, but not in most other organic solvents. Its empirical formula is $C_{63}H_{88}O_{14}N_{14}PCo$, giving it a cobalt content of 4·35 %. The central cobalt atom is in the trivalent state, but is easily reduced to the divalent condition. Vitamin B_{12} contains four pyrrole rings, as does heme, but these rings are partially reduced whereas in heme they are fully conjugated. Also, in heme, the groups binding the pyrrole rings together are all alike, but in vitamin B_{12} one bridging —CH— group is missing, so there is a five-membered ring in place of a six-membered ring. There is also a direct cobalt–carbon bond. Vitamin B_{12} brings about molecular re-arrangements, moving an organic group from one carbon atom in the substrate to another.

Vitamin B_{12} is the first vitamin found to contain a metal, the only cobalt-containing compound in the human body, and the most complex non-polymer yet found in nature. It is also one of the most physiologically potent compounds; only about 1 μg per day is required in human nutrition. It is essential, in conjunction with an enzyme, in at least ten reactions, although only one of these is present in man, where it affects growth and red blood cell formation.

Vitamin B_{12} is the only vitamin that is synthesized exclusively by microorganisms. Some bacteria and protozoa cannot synthesize the vitamin; other bacteria and actinomycetes make far more than they need, and are actually employed for the manufacture of the vitamin.[8] The most concentrated natural synthesis occurs in the fore-stomach of ruminants. Bacterial synthesis of the vitamin also takes place in the gut of other species, including humans. In some animals and in man, however, the synthesis occurs too low down in the gastro-intestinal tract for the vitamin to be absorbed; an external dietary source is required.

Exactly how vitamin B_{12} works is still a mystery. In nature, vitamin B_{12} always functions as a co-enzyme along with an enzyme to catalyze a reaction. In the latter, it is very difficult to differentiate what the enzyme does and what vitamin B_{12} does.

Cobalt in Human Foods

Cobalt proceeds from soil and water into plants and animals, and eventually to humans. In Table 4, Chapter 7, data are tabulated on the cobalt content of a large number of plants and plant products, many of which are common in human nutrition.

Foods of plant origin containing a high cobalt content, about 0.2 mg kg^{-1} or over, include beet greens, bread, buckwheat, cabbage, figs, green onions, molasses, mushrooms, pears, radishes, spinach, tomato, and turnip greens. Low cobalt foods, containing less than about 0.05 mg kg^{-1}, include apples, apricots, bananas, carrots, cassava, cherries, coffee, corn, eggplant, oats, pepper, potatoes, rice, salt, sweet potatoes, wheat, and yams. The following may be listed as intermediate in cobalt content: barley, beetroot, chard, horseradish, peas, rye, strawberry, walnut, water cress, watermelon, and wax beans. It is worthy of note that cassava, corn, potatoes, rice, and wheat—the staple foods for so many humans—are usually low in cobalt.

The cobalt contents of a number of animal products and fish used in human nutrition are given in Table 7. It will be observed that most values reported for cobalt in the livers of beef, chicken, cod, lamb, pig, and sheep are considerably higher than in other parts of the animal, bird, or fish.

Table 7   Cobalt Content of Animal Products and Fish

	Cobalt (mg kg^{-1}) dry basis	Ref.
Animal meat	0.03–0.09	9
Animal products	0.006–0.065	10
Beef	0.011	11
Beef liver	0.076	12
	0.135	13
	0.201	14
	0.127	15
	0.145	16
	0.35	17
Chicken liver	0.12	18

Table 7 (cont'd)

	Cobalt (mg kg^{-1}) dry basis	Ref.
Chicken meat	0·0057	19
Cured meat	0–0·41	20
Lamb liver	0·125	16
Meat products	0·012–0·15	21
Pig liver	0·10	16
	0·42–1·18	22
Sheep liver	0·04–0·05	23
	0·112	12
	0·148	13
	0·43	24
	0·6–1·0	25
Veal	0·003	11
Egg white	0·034	17
Egg yolk	0·068	17
Eggs, shelled	0·03	26
Milk	0·007	16
	0·0065–0·0395	21
	0·017	9
	0·00084	27
	0·010	13
	0·082–0·208	28
	0·016	12
	0·034–0·041	29
	0·072–0·124	30
Ocean fish	0·1	31
Freshwater fish	0·09	31
Cod muscle	0·0021	32
Cod liver	0·0111	32
Carp muscle	<0·001	32
Whiting	0·28	33
Osmerus eperlanus	1·1	33
Gadus aeglefinus	0·1	33
Gadus morrhua	0·07	33
Gadus virens	0·07	33
Mollusks	0·01	31
Crustaceans	0·04	31

The reported cobalt contents of the same human foods vary widely in many instances. This is, of course, to be expected in view of widely different soil and climatic conditions throughout the world. The published information on fish is meager in this field, and it will be interesting to see, when more data are available, whether similar differences will be found in the cobalt contents of the same fish species. A number of years ago, it was observed that the amount of cobalt in the tissues of fish appeared to be higher than that of nickel, whereas in sea water the reverse is true.[33] No verification or refutation of this statement has appeared; such a study would be of great interest in ichthyology.

Three wines from the Tashkent region of Russia had 0·021–0028 mg cobalt l^{-1}.[34] Another intriguing item of information on cobalt in human nutrition is the report that kumiss, the intoxicating liquor made from mares' milk, contained 2·7–16 mg cobalt kg^{-1} of dry matter.[35]

The cobalt content of a wide variety of fresh, frozen, and canned fruit and vegetables ranged from 0·01–0·15 mg kg^{-1}; the mean level in both fresh and canned foods was 0·2 mg kg^{-1}.[36] Nuts were reported to have a high concentration of cobalt, and mushrooms contained the maximum amount of cobalt, at 0·61 mg kg^{-1}, in a children's institution in Ivanov.[37] A publication has given good sources of cobalt for humans, >1 mg kg^{-1}, as fish, cocoa, bran, and molasses.[38]

Cobalt Intake in Humans

Daily rations in the city of Ivano-Frankovsk were estimated to contain 0·1618–0·3238 mg cobalt kg^{-1}.[15] Another publication puts the daily intake of cobalt at 0·140–0·580 mg.[38] In a study of Transcarpathian foods, more than half of the daily diet of 0·0194–0·0237 mg cobalt was obtained from grain and grain products, 12–15% came from animal products, and the remainder was obtained from vegetables and fruits.[10]

The average uptake of cobalt for humans has been placed at 0·2–0·3 mg per day;[39] other papers estimate the daily adult intake at 0·03 mg cobalt,[40] and state that a positive balance is observed with a daily diet of 0·03 mg cobalt.[41] The diet of children 2–6 years of age contains 0·06397–0·07146 mg cobalt per day.[42] The cobalt content of average diets of various age groups in the Netherlands has been listed.[43] Two-thirds of Japanese adults take 0·011–0·028 mg cobalt daily; rural children consume half this amount, or less.[44] A daily diet containing 0·036 mg cobalt has been stated to fully cover requirements.[45] For pre-adolescent girls, a suggested daily allowance of cobalt is 0·015 mg.[46]

Cobalt in Human Blood and Tissues

A number of papers report the cobalt content of blood and body tissues of humans. In erythrocytes, cobalt ranged from 0·059–0·13 mg kg^{-1};[47] in blood serum, 0.0055–0.029 mg kg^{-1},[47] 0.061–0.063 mg kg^{-1},[48] 0.05–0.40 mg kg^{-1},[49] 0.0066 mg kg^{-1},[50,51] and in whole blood, 0.238 mg kg^{-1}.[48]

In healthy children of school age, the ratio of cobalt to nickel in the blood was given as $1:3.7$.[52] The amount of cobalt in the blood of subjects aged 50–100 years is lower than in those aged 20–50 years; blood cobalt is somewhat higher in men than in women at all ages.[53] The cobalt content in the blood of normal persons was found to be highest in August and lowest in January; this was related to the maximum cobalt intake in vegetables and milk-products being in May–July, and the minimum in January–February.[54]

Healthy children of 6–7 years, on an adequate diet, averaged 0.00075 mg cobalt kg^{-1} body weight; of an average daily intake of 0.040–0.042 mg cobalt, 0.002–0.003 mg was excreted in urine, and 0.035–0.038 mg in the feces.[55] Cobalt in adult tissues ranged from a mean value of 0.0003 mg kg^{-1} in serum to 0.07 mg kg^{-1} wet weight in liver; the livers of newborn infants contained much lower concentrations than those of adults.[56]

The cobalt content of menstrual blood of women of various ages ranged from 0.2375–0.5565 mg kg^{-1}; the cobalt content of venous blood of non-pregnant women was 0.08–0.16 mg kg^{-1}.[57] A correlation was found between the cobalt level in the blood of pregnant and delivering women, and the cobalt content in the umbilical cord; placentas from normal births contained more cobalt than did the blood.[58] The highest level of cobalt was found in the non-pregnant woman, and the lowest was at the time of childbirth; during lactation, blood cobalt increased to 0.0749 mg kg^{-1}.[59]

In normal human dentine, the cobalt content has been reported as 0.00034 mg kg^{-1}.[60] It was found that female hair contained significantly more cobalt than male hair; the cobalt content did not decline with age.[61] In individual parts of the human eye, cobalt was found in the ash to the extent of 0.167–0.362%; cobalt was considerably increased in eyes enucleated by glaucoma.[62]

The cobalt content of human skin varied from 0.092–0.114 mg kg^{-1}; it decreased with age, and was not dependent on sex.[63] The cobalt in human myocardium is combined with that portion which can be extracted with acetic acid at pH 4.[64] The cobalt content in the uterine mucous membrane decreased with age and increased with the number of pregnancies.[65] One worker reported that 22.5% of the total intake of 0.036 mg cobalt was excreted in urine,[45] while another paper reported that about 12% of the total daily intake of 0.051 mg cobalt was excreted in urine.[66]

It has been found that the cobalt content of human embryos increases with age.[67] The average cobalt contents in ovaries of infants born after 28 and 40 weeks of pregnancy were 13.69 and 6.89 mg kg^{-1}, respectively.[68] In myocardium of the human fetus, cobalt decreased from 12–19 weeks to birth.[69] There is no accumulation of cobalt in the human liver before birth.[70]

The amount of cobalt in human milk decreases continuously with the period of lactation, from 0.008 mg l^{-1} to 0.[71] It has been found that cobalt in colostrum, transitional, and mature breast milk was, in mg l^{-1}, trace to 0.032, $0.004–0.044$, and $0.004–0.016$, respectively.[72] Another investigator reported the cobalt content of human milk as $0.006–0.023$ mg l^{-1}, and that of placenta as $0.0105–0.024$ mg kg^{-1}.[73] A paper stated that in the first three months, breast-fed infants received 0.00138 mg cobalt kg^{-1} body weight; with breast feeding, 23% of the cobalt intake was excreted in the feces and 8% in the urine.[74]

Cobalt in Human Biochemistry and Physiology

Cobalt and iron appear to share a similar absorptive mechanism in the proximal intestine, but not in the distal.[75–77]

The absorption of cobalt occurs at sites in the gastro-intestinal tract above the region where protein digestion takes place.[78] The retention of cobalt is increased when antibiotics are given, and also when dietary copper increases.[79] In pancreatic islet tissue, the β cells and possibly the α_1 cells concentrate cobalt, while the α_2 cells and the agranular parenchymal cells show no affinity for cobalt; this might be related to the high power of cobalt to crystallize insulin.[80]

In a study of the effect of cobalt salts on human cells in tissue cultures, it appeared that ribonucleic acid, deoxyribonucleic acid, and chondroitin sulphate will react with complex cobalt salts.[81] The connection between alimentary anemia and a deficiency of cobalt and copper has been confirmed; cobalt greatly influenced the number of erythrocytes, and copper affected the hemoglobin concentration.[82] Daily oral administration of 150 mg cobalt chloride to 10 people, for 8–10 days, appeared to stimulate erythropoietic activity.[83]

Cobalt chloride in human saliva increased amylolytic activity, but neither the nitrate nor sulphate of cobalt had any significant effect. Cobalt chloride, and the nitrate and sulphate in larger doses, increased the activity of pancreatic amylase and lipase.[84] The addition of cobalt chloride to a suspension of human erythrocytes slightly decreased the affinity of hemoglobin for carbon monoxide.[85] Cobalt-induced cardiac insufficiency and other side effects of cobalt therapy in the treatment of anemia have been reviewed.[86] A single dose of cobalt chloride increased blood serum triglycerides, free fatty acids, and total glycerol.[87]

Cobalt inhibited the mineralization of rachitic cartilage when present at a concentration of 0.1 M or less; cobalt was unique in preventing apatite crystal formation at concentrations inhibiting mineralization.[88] Cobalt did not impair the efficiency of rat liver mitochondria oxidative phosphorylation during brief periods in which mitochondrial damage was at a minimum.[89] Vaccinia virus was inactivated by cobalt.[90] A solution containing 0·005 mg cobalt chloride per liter added to a monolayer culture of fetal rat dermis fibroblasts killed nearly all the latter in 48 h.[91] Cobalt caused enlargement of the cell vacuole of *Cricosphaera carterae*, and the appearance of membrane-bound vacuoles containing electron-dense bodies.[92]

Several papers have appeared on the antidotal properties of cobalt against certain poisons. In experimental aniline and lead poisoning, saprogel and cobalt gave the best recovery.[93] Cobalt has a marked tendency to form various complexes with hydrocyanic acid, such as $K_3(Co(CN)_6)$, and to detoxify cyanide.[94] Against cyanide intoxication, cobalt EDTA is a more effective antidote than the sodium nitrite-hyposulphite complex.[95]

It was observed many years ago that food products from places known for endemic goiter usually had low cobalt contents.[96-99] The application of cobalt, even in the absence of iodine, seemed to have a beneficial effect on the incidence of goiter. Investigators have found that cobalt apparently counteracts iodine deficiency without changing the weight of the glands,[100] and that cobalt activates thyroid gland activity in the case of an iodine insufficiency, but best results are obtained by the combined use of cobalt and iodine.[101]

Finally, a few individual studies may be cited on the biochemical behavior of cobalt. Cobalt complexes with some proteins and polypeptides have been described.[102] A cobalt–histidine complex was found to have no oxygen atoms of the oxonium type, but did contain a peroxide bridge between two six-coordinated cobalt ions.[103] In a study of the cobalt-treated single node of Ranvier, it was postulated that the inactivation of the sodium carrier system was inhibited by cobalt, and that the latter selectively inhibited potassium permeability.[104] It was found that the incorporation of leucine into soluble ribonucleic acid was greatly increased in the presence of cobalt, and the latter could not be replaced by any other metal.[105] It has been stated that cobalt may act at an intracellular site, preventing the physiological utilization of calcium in the excitation-contraction coupling system of the isolated smooth muscle.[106]

Toxicity of Cobalt to Humans

Studies on cobalt sensitization[107,108] have shown that sensitivity to cobalt should be considered proven if the epicutaneous test is positive after 72 h, with a 2% solution of cobalt, the intracutaneous test is positive 72 h later in 1:1000 dilution, and contact with cobalt has been established.

In 1966, a previous regulation under the U.S. Federal Food, Drug, and Cosmetic Act, providing for the use of cobaltous salts in fermented malt beverages, was revoked.[109] Prior to this, up to 1.2 parts 10^{-6} cobalt, in the form of either the acetate, chloride, or sulphate of cobalt, could be used to improve foam stability and prevent gushing. Certain disorders associated with a heavy consumption of beer led to the ban on the use of cobalt in these beverages.

Some investigations have been made on the toxicity of dusts from the production and use of cemented carbides. The latter, usually tungsten carbides, are commonly bonded with cobalt powder. The machining operations of grinding, cutting, sawing, drilling, etc. in which the cemented carbides are employed, give rise to fine dusts or aerosols which have been of occasional concern to those in the field of industrial hygiene. A British study of hard metal, or tungsten carbide, disease found the incidence to be 1 in 255 workers, and stated that pulmonary response may be due to sensitivity. The investigators also reported that cobalt may be the toxic agent, though this metal has not been found in post-mortem analyses of lung tissue, perhaps because of its high solubility in plasma.[110]

Lung tissue has been found to contain tungsten, titanium, and cobalt.[111] Increased blood globulins, sometimes with increased hemoglobin, can be attributed to cobalt, but the cause of the lung changes is still uncertain.

In the dust from the production of sintered carbides, the increased toxicity of cobalt in cobalt–tungsten carbide mixtures was ascribed to the higher solubility of cobalt in contact with tungsten.[112]

During wet-process tungsten carbide grinding, air sampling gave an average cobalt concentration of $0.01–0.02$ mg m^{-3} in the coolant mist aerosols containing carbide dust which would be inhaled.[113]

The possible toxic effects of other cobalt compounds have been studied. A clinical investigation of 247 workers experiencing varying degrees of exposure to cobalt compounds showed the following: digestive troubles, 25; reticulocytosis, 17; anemia, 25; catarrhal rhinitis, 23; catarrhal laryngopharyngitis, 60; chronic bronchitis, 35; pneumosclerosis, 33; gastritis, 17; and hypotonia, 52.[114] The trouble with reports of this type is that there is no indication of whether or not the workers were currently, or had been previously, exposed to other chemical compounds.

One group has examined persons processing an iron–nickel ore containing 0.6% cobalt, and exposed to $0.063–1.605$ mg m^{-3} cobalt, in the form of aerosols.[115] The following percentage increases over unexposed subjects were found: seromucoids, 5.5; sialic acid, 21.9; and hexosamine, 23.2%.

It has been pointed out that the effect of cobalt on the body is of a temporary nature. The effects on the thyroid gland, central nervous system, and renal tubules were completely reversible after administration of the drug was stopped. Cobalt chloride is very effective in raising the level of hemoglobin in refractory anemias, especially that owing to renal failure.[116]

In countries enjoying a reasonable standard of living, human foods are diversified and usually come from several or many areas. There is, consequently, far less chance for man to suffer from a cobalt deficiency than for ruminants. In primitive communities, however, where nearly all food is obtained from crops and animals raised in the immediate vicinity, there is a possibility of cobalt deficiency. It is not inconceivable that certain nutritional disorders in parts of Africa, Asia, or South America may eventually be linked with a low cobalt intake.[8]

Apart from the natural cobalt content of foodstuffs and waters, minute quantities of this element may be inadvertently added to human food in a variety of unsuspected ways. Cobalt is an important substance for producing a blue color in glass, pottery, porcelain, and china, or for neutralizing the yellow tint of iron compounds in ceramic materials. The adhesion of enamel to steel is greatly improved by cobalt, and the latter is widely employed as a ground coat for kitchen enamelware. Nickel, Monel, or stainless steel, which contain traces of cobalt, are frequently found in food processing operations and household kitchens. Many driers for paints, varnishes, and lacquers are cobalt-containing compounds. Trace amounts of cobalt may participate in human nutrition as contaminants or dusts from the production or operation of cobalt-containing alloys such as magnets, high-temperature and wear-resistant alloys, cemented carbides, high-speed steel, nickel electroplating, and from cobalt-containing catalysts.

Reviews

Many aspects of cobalt in human nutrition have been reviewed in books and technical journals. The subject of vitamin B_{12} has already been cited.[6,7] Cobalt in foods, generally, has been reviewed in two papers,[117,118] and cobalt in nutrition in another.[119] Compilations of the cobalt content of a number of foods are given in two papers;[117,120] the subject is also discussed in a biomedical book.[121]

The metabolism of cobalt,[122] and the metabolic features of cobalamin deficiency in man[123] have been surveyed; a book on mineral metabolism contains a chapter on cobalt.[124]

Papers have appeared, reviewing the biochemical function of cobalt,[125] its role and significance for living organisms,[126] and the physical–chemical behavior of cobalt in hydro-biological systems.[127]

A selective survey of cobalt and other trace elements in clinical chemistry forms the subject of a review paper.[128] Two books discuss, respectively, cobalt and other trace elements in human and animal nutrition,[129] and the soil factor in human and animal nutrition.[130]

References

1. Shohl, A. T. (1939). "Mineral Metabolism". Reinhold, New York.
2. Sherman, H. C. (1952). "Chemistry of Food and Nutrition", 8th Edn. Macmillan, New York.
3. McCollum, E. V., Orent-Keiles, E., and Day, H. G. (1944). "The Newer Knowledge of Nutrition". Macmillan, New York.
4. Smith, E. L. (1948). *Nature, Lond.* **162**, 144–5.
5. Rickes, E. L., Brink, N. S., Koniuszy, F. R., Wood, T. R., and Folkers, K. (1948). *Science* **107**, 396–7.
6. Smith, E. L. (1965). "Vitamin B_{12}", 3rd Edn. Wiley, New York.
7. Hodgkin, D. C. (1969). *Proc. R. Inst. Gt Brit.* **42**, Pt. 6, 377–96.
8. Young, R. S. (1960). "Cobalt", Am. Chem. Soc. Monograph 149. Reinhold, New York.
9. Goto, T. (1954–5). *J. Japan. Soc. Fd Nutr.* **7**, 102–3; (1959). *C.A.* **53**, 7449.
10. Kotelyanskaya, L. I. (1963). *Vop. Pitan.* **22**, 71–2; (1964). *C.A.* **60**, 7359.
11. Hecht, H. (1973). *Archiv. Lebensmittelhyg.* **24**, 255–8; (1974). *C.A.* **80**, 144458.
12. Berestova, V. I. and Panova, M. K. (1965). *Uchen. Zap. Petrozavodsk. Gos. Univ.* **13**, 107–9; (1966). *C.A.* **65**, 19209.
13. Meleshko, K. V. (1959). *Vop. Pitan.* **18**, 57–61; (1959). *C.A.* **53**, 18319.
14. Correa, R. (1957). *Arq. Inst. Biol.* **24**, 199–227; (1960). *C.A.* **54**, 21359.
15. Soroka, N. V. (1966). *Vop. Pitan.* **25**, 80–3; (1966). *C.A.* **65**, 14327.
16. Taktakishvili, S. D. (1963). *Vop. Pitan.* **22**, 73–4; (1964). *C.A.* **60**, 7359.
17. Skoropostizhnaya, A. S. (1957). *Vop. Pitan.* **16**, 59–62; (1957). *C.A.* **51**, 1599.
18. Maturova, E. T. (1964). Mikroelem. Biosfere Ikh Primen. Sel'. Khoz. Med. Sib. Dal'nego Vostoka, Dokl. Sib. Konf. 2nd, 501–2; (1964). *C.A.* **70**, 85390.
19. Khamidullina, L. (1975). *Vop. Pitan.* No. 4, 79–80; (1975). *C.A.* **83**, 162318.
20. Kirkpatrick, D. C. and Coffin, D. E. (1975). *J. Sci. Food Agric.* **26**, 43–6.
21. Grigor'yants, N. N. (1967). *Izv. Akad. Nauk Turkmen. S.S.R., Ser. Biol. Nauk* No. 4, 45–8; (1967). *C.A.* **67**, 115825.
22. Taucins, E. and Svilane, A. (1965). *Fiziol. Zhivotn. Inst. Biol. Akad. Nauk Latv. S.S.R.* **4**, 251–4; (1967). *C.A.* **66**, 17385.
23. Euybov, I. Z. (1967). *Veterinariya* **44**, 98–100; (1968). *C.A.* **68**, 1220.
24. Kazaryan, E. S. and Airuni, G. A. (1967). *Izv. Sel'.-khoz. Nauk Min. Sel'. Khoz. Arm. S.S.R.* **10**, 65–72; (1968). *C.A.* **68**, 28789.
25. Rish, M. A., Ben-Utyaeva, G. S., and Shimanov, V. G. (1958). *Nauch Tr. Nauch.-Issled. Inst Karakulevod. Uzbek. Akad. Sel'.-khoz. Nauk* **7**, 249–61; (1959). *C.A.* **53**, 22361.
26. Kirkpatrick, D. C. and Coffin, D. E. (1975). *J. Sci. Food Agric.* **26**, 99–103.
27. Kiermeier, F. and Winkelman, H. (1961). *Z. Lebensmitterlunters. u.-Forsch.* **115**, 309–22; (1962). *C.A.* **56**, 742.
28. Leonov, V. A., Terent'eva, M. V., and Gorski, N. A. (1960). *Vest. Akad. Navuk Belaruss. S.S.R., Ser. Biyal. Navuk* No. 3, 47–55; (1962). *C.A.* **56**, 15843.
29. Odynets, R. N. and Valuiskii, P. P. (1959). *Izv. Akad. Nauk Kirg. S.S.R.* 1, *Ser. Biol. Nauk* No. 1, 127–38; (1960). *C.A.* **54**, 15569.
30. Vsyakikh, M. I. (1959). Int. Dairy Cong. Proc. 15th Cong. London, Vol. 3, 1761–5; (1960). *C.A.* **54**, 12411.
31. Mershina, K. M., Khalina, N. M., and Krasnitskaya, A. I. (1962). Mikroelementy v Vost. Sib. i na Dal'n. Vost. Inform. Byal. Koordinats. Komis.

po Mikroelementam dlya Sib. i Dal'n. Vost. No. 1, 35–43; (1963). *C.A.* **59**, 13270.

32. Chojnicka, B. and Szyszko, E. (1964). *Rocz. Pánst. Zakl. Hig.* **15**, 23–6; (1964). *C.A.* **61**, 6275.
33. Vinogradov, A. P. (1953). "The Elementary Chemical Composition of Marine Organisms". Sears Foundation for Marine Research, New Haven.
34. Kruglova, E. K., Berezhkovskii, E. A., and Shapiro, L. V. (1965). Mikroelementy v Sel'. Khoz., Akad. Nauk Uzbek. S.S.R., Otdel. Khim. Tekhnol. i Biol. Nauk, 423–36; (1966). *C.A.* **64**, 13331.
35. Kiryukhin, R. A. (1959). *Konevodstvo* No. 7, 31–2; (1963). *C.A.* **59**, 4318.
36. Thomas, B., Roughan, J. A., and Watters, E. D. (1974). *J. Sci. Food Agric.* **25**, 771–6.
37. Lago, O. M. (1965). *Sb. Nauch. Tr. Ivanov. Gos. Med. Inst.* No. 31, 78–84; (1967). *C.A.* **67**, 71479.
38. Schroeder, H. A., Nason, A. P., and Tipton, I. H. (1967). *J. Chron. Dis.* **20**, 869–90.
39. Kolomiitseva, M. G. (1966). *Gig. Pitan.* 121–6; (1969). *C.A.* **70**, 17965.
40. Ouren, I. (1957). Univ. Bergen Arbok Naturvitenskap. Rekke, No. 9; (1959). *C.A.* **53**, 6460.
41. Nodiya, P. I. (1972). *Gig. Sanit.* **37**, 108–9; (1972). *C.A.* **77**, 73946.
42. Vorob'eva, A. I. and Osmolovskaya, E. V. (1970). *Gig. Sanit.* **35**, 108–9; (1971). *C.A.* **74**, 51177.
43. Belz, R. (1960). *Voeding* **21**, 236–51; (1960). *C.A.* **54**, 25098.
44. Yamagata, N., Kurioka, W., and Shimizu, T. (1963). *J. Radiation Res.* **4**, 8–15; (1964). *C.A.* **61**, 7430.
45. Taktakishvili, S. D. (1963). Sb. Tr. Nauch.-Issled. Inst. Sanit. i Gig. Gruz S.S.R., Tbilisi, 213–17; (1965). *C.A.* **62**, 10948.
46. Engel, R. W., Price, N. O., and Miller, R. F. (1967). *J. Nutr.* **92**, 197–204.
47. Idel'son, L. I. and Zhukovskaya, E. D. (1963). *Probl. Gematol. Pereliv. Krovi* **8**, 10–14; (1963). *C.A.* **59**, 6787.
48. Butt, E. M., Nusbaum, R. E., Gilmour, T. C., Didio, S. L., and Mariano, Sister (1964). *Archs Environ. Hlth* **8**, 52–7.
49. Mukhamedova, I. G. and Mallina, E. R. (1966). *Med. Zh. Uzbek.* No. 10, 33; (1967). *C.A.* **66**, 113680.
50. Mertz, D. P., Wilk, G., and Koschnick, R. (1968). *Verh. Dt. Ges. Inn. Med.* **74**, 600–3; (1969). *C.A.* **71**, 28350.
51. Mertz, D. P., Koschnick, R., Wilk, G., and Pfeilsticker, K. (1968). *Z. Klin. Chem. Klin. Biochem.* **6**, 171–4; (1968). *C.A.* **69**, 17321.
52. Tyurina, N. S. (1964). Materialy Nauch. Konf. Chelyab. Med. Inst. Chelyabinsk Sb., 306–8; (1965). *C.A.* **63**, 15304.
53. Korobenkova, M. M. (1962). *Dokl. Akad. Nauk Beloruss. S.S.R.* **6**, 385–6; (1962). *C.A.* **57**, 11698.
54. Chakhovskii, I. A. (1961). *Zdravookhr. Belorussii* No. 6, 17–21; (1962). *C.A.* **57**, 2646.
55. Ripak, E. N. (1961). *Vop. Pitan.* **20**, 19–22; (1962). *C.A.* **57**, 17161.
56. Parr, R. M. and Taylor, D. M. (1964). *Biochem. J.* **91**, 424–31.
57. Osadchaya, O. V. (1962). *Zdravookhr. Belorussii* No. 8, 48–51; (1963). *C.A.* **59**, 9157.
58. Asmolovs'kii, G. V. (1959). *Nauk. Zap. Stanislovs'k. Med. Inst.* No. 3, 193–200; (1963). *C.A.* **59**, 10558.

59. Agranovskaya, B. A. (1966). *Akush. Ginekol.* **42**, 34–7; (1967). *C.A.* **66**, 17466.
60. Soremark, R. and Lunsberg, M. (1964). *Odontol. Revy* **15**, 285–9; (1965). *C.A.* **62**, 4407.
61. Schroeder, H. A. and Nason, A. P. (1969). *J. Invest. Dermat.* **53**, 71–8.
62. Shlopak, I. V. (1962). Mikroelementy v Sel'. Khoz. i Med., Ukr. Nauch.-Issled. Inst. Fiziol. Rast. Akad. Nauk Ukr. S.S.R., Materialy 4-go (Chetvertogo) Vses. Soveshch. Kiev, 632–5; (1965). *C.A.* **63**, 10401.
63. Yagovdik, N. Z. (1964). *Vest. Akad. Nauk Belaruss. S.S.R., Ser. Biyal. Navuk* No. 3, 93–7; (1965). *C.A.* **62**, 12240.
64. Kneta, Z. (1963). *Latv. PSR Zināt. Akad. Vest.* No. 11, 107–14; (1964). *C.A.* **60**, 12468.
65. Osadchaya, O. V. (1963). *Zhravookhr. Belorussii* No. 6, 43–4; (1964). *C.A.* **61**, 9838.
66. Idel'son, L. I. and Zhukovskaya, E. D. (1963). *Probl. Gematol. Pereliv. Krovi* **8**, 24–6; (1964). *C.A.* **60**, 3344.
67. Sakovich, L. T. (1957). *Sb. Nauch Rabot., Minsk. Med. Inst.* **19**, 259–67; (1959). *C.A.* **53**, 10436.
68. Osadchaya, O. V. (1963). *Zdravookhr. Belorussii* **12**, 35–6; (1964). *C.A.* **61**, 15116.
69. Krasnoshlykov, G. Ya. (1965). Trudȳ 5-i Nauchn. Sessii. Aktyubinsk Med. Inst., 73–5; (1967). *C.A.* **67**, 88772.
70. Widdowson, E. M., Chance, H., Harrison, G. E., and Milner, R. D. G. (1972). *Biol. Neonate* **20**, 360–7.
71. Leonov, V. A. (1958). *Vest. Akad. Navuk Belaruss. S.S.R., Ser. Biyal. Navuk* No. 3, 69–74; (1960). *C.A.* **54**, 19891.
72. Kopytkova, O. G. (1959). *Izv. Akad. Nauk Beloruss. S.S.R., Ser. Biol. Nauk* No. 4, 142–4; (1961). *C.A.* **55**, 5705.
73. Ozols, A. (1960). *Trudȳ Inst. Eks. Med., Akad. Nauk Latv. S.S.R.* **22**, 129–34; (1961). *C.A.* **55**, 11589.
74. Reshetkina, L. P. and Orinchak, M. A. (1974). *Pediat. Akush. Ginekol.* No. 3, 24–6; (1975). *C.A.* **83**, 162453.
75. Schade, S. G., Felsher, B. F., and Bernier, G. M. (1970). *J. Lab. Clin. Med.* **75**, 435–41.
76. Valberg, L. S. (1971). Intestinal Absorption Metal Ions, Trace Elem. Radionuclides, 257–63.
77. Thomson, A. B. R., Shaver, C., Lee, D. J., Jones, B. L., and Valberg, L. S. (1971). *Am. J. Physiol.* **220**, 674–8.
78. Paley, K. R. and Sussman, E. S. (1963). *Metabolism* **12**, 975–82.
79. Kirchgessner, M. (1965). *Proc. Nutr. Soc.* **24**, 88–99.
80. Falkner, S., Knutson, F., and Voigt, G. E. (1964). *Diabetes* **13**, 400–7; (1965). *C.A.* **62**, 2081.
81. Lazzarini, A. A. and Weissmann, G. (1960). *Science* **131**, 1736–7.
82. Kolomiitseva, M. G. (1961). *Probl. Gematol. Pereliv. Krovi* **6**, 38–41; (1961). *C.A.* **55**, 26158.
83. Panayotopoulos, E., Valtis, D., Concouris, L., and Goulis, G. (1961). Proc. Cong. European Soc. Haematol. 8th, Vienna, Art. 481; (1963). *C.A.* **58**, 2710.
84. Shkol'nik, M. I. (1965). *Uchen. Zap. Petrozavodsk. Gos. Univ.* **12**, 136–40; (1966). *C.A.* **65**, 2559.
85. Paulet, G. and Chevrier, R. (1966). *C.R. Soc. Biol.* **160**, 1726–7; (1967). *C.A.* **66**, 103117.

86. Achenbach, H. and Urbaszek, W. (1972). *Z. Ges. Inn. Med. Grenzgeb.* **27**, 809–16; (1973). *C.A.* **78**, 80403.
87. Munoz-Calvo, R., Valcazar, A., and Lucas, J. (1973). *Revta Espan. Fisiol.* **29**, 61–4; (1974). *C.A.* **80**, 10924.
88. Bird, E. D. and Thomas, W. C. (1963). *Proc. Soc. Exp. Biol. Med.* **112**, 640–3.
89. Strickland, E. H. and Goucher, C. R. (1963). *Nature, Lond.* **198**, 790–1.
90. Kaplan, C. (1963). *J. Gen. Microbiol.* **31**, 311–14.
91. Daniel, M. R., Dingle, J. T., Webb, M., and Heath, J. C. (1963). *Brit. J. Exp. Pathol.* **44**, 163–76.
92. Blankenship, M. L. and Wilbur, K. M. (1975). *J. Phycol.* **11**, 211–19.
93. Lescinskaite, A. (1960). *Liet. TSR Mokslų. Akad. Darb., Ser. C.* No. 2, 163–74; (1961). *C.A.* **55**, 10618.
94. Bartelheimer, E. W., Friedberg, K. D., and Lendle, L. (1962). *Archs Intern. Pharmacodyn.* **139**, 99–108; (1963). *C.A.* **58**, 3811.
95. Terzic, M. and Milosevic, M. (1963). *Therapie* **18**, 55–61; (1965). *C.A.* **62**, 5791.
96. Novikova, E. P. (1965). Mikroelementy v Zhivotnovod. i Med. Akad. Nauk Ukr. S.S.R. Resp. Mezhvedomstv. Sb., 57–61; (1966). *C.A.* **64**, 13283.
97. Shalaev, F. T. (1960). *Izv. Akad. Nauk Kirg. S.S.R., Ser. Biol. Nauk* **2**, 85–93; (1961). *C.A.* **55**, 26337.
98. Novikova, E. P. (1961). *Gig. Sanit.* **29**, 80–2; (1962). *C.A.* **56**, 7059.
99. Gurevich, G. P. and Malyutina, L. I. (1962). *Trudÿ Vladivostoksk. Nauch.-Issled. Inst. Epidemiol. Mikrobiol. Gig., Sb.* No. 2, 211–13; (1963). *C.A.* **59**, 14519.
100. Koval'skii, V. V. and Blokhina, R. I. (1963). *Probl. Endokrinol. Gormonoterap.* **9**, 42–6; (1963). *C.A.* **60**, 8396.
101. Koval'skii, V. V. and Blokhina, R. I. (1962). Mikroelementy v Sel'. Khoz. i Med., Ukr. Nauch.-Issled. Inst. Fiziol. Rast., Akad. Nauk Ukr. S.S.R., Materialy 4-go (Chetvertogo) Vses. Soveshch. Kiev, 486–90; (1965). *C.A.* **63**, 8796.
102. Bello, J. and Bello, H. R. (1961). *Nature, Lond.* **192**, 1184–5.
103. Sano, Y. and Tanable, H. (1963). *J. Morg. Nucl. Chem.* **25**, 11–15.
104. Hashimura, S. and Osa, T. (1963). *Japan J. Physiol.* **13**, 219–30; (1963). *C.A.* **59**, 11970.
105. Devi, A. and Sarkar, N. K. (1963). *Biochim. Biophys. Acta* **68**, 254–62.
106. De Moraes, S. and Carvalho, F. V. (1969). *Pharmacology* **2**, 230–6.
107. Valer, M., Somogyi, Z., and Racz, I. (1966). *Börgyög. Venerol. Szemle* **42**, 1–12; (1966). *C.A.* **65**, 11223.
108. Valer, M., Somogyi, Z., and Racz, I. (1967). *Dermatologica* **134**, 36–50; (1967). *C.A.* **66**, 79330.
109. Anon. (1966). *Federal Register* **31**, 10744, Aug. 12; (1966). *C.A.* **65**, 12771.
110. Bech, A. D., Kipling, M. D., and Heather, J. C. (1962). *Brit. J. Ind. Med.* **19**, 239–52.
111. Kipling, M. D. (1963). Int. Cong. Occupational Health, 14, Madrid, Vol. 2, 680–1; (1966). *C.A.* **64**, 8843.
112. Kaplun, Z. S. and Mezentseva, N. V. (1963). Toksikol. Redkikh Metal., 227–38; (1964). *C.A.* **60**, 4680.
113. Lichtenstein, M. E., Bartl, F., and Pierce, R. T. (1975). *J. m. Ind. Hyg. Assoc.* **36**, 879–85.
114. Kaplun, Z. S. (1963). Toksikol. Redkikh Metal., 164–76; (1964). *C.A.* **60**, 2246.
115. Klucik, I., Kemka, R., and Vladar, M. (1967). *Bratislav. Lék. Listy* **48**, 355–65; (1967). *C.A.* **67**, 36157.
116. Schirrmacher, U. O. E. (1967). *Brit. Med. J.* **1**, 544–5; (1967). *C.A.* **66**, 93552.

117. Schlettwein-Gsell, D. (1970). *Int. Z. Vitaminforsch.* **40**, 673–83; (1971). *C.A.* **74**, 63205.
118. Schlettwein-Gsell, D. and Mommsen-Straub, S. (1973). *Int. J. Vitam. Nutr. Res., Beih.* **13**, 188 pp; (1976). *C.A.* **85**, 33918.
119. Underwood, E. J. (1975). *Nutr. Rev.* **33**, 65–9.
120. Clemente, G. F. (1976). *J. Radioanal. Chem.* **32**, 25–41.
121. Waslien, C. I. (1976). *In* "Trace Elements in Human Health and Disease" (A. S. Prasad, ed.), Vol. 2. Academic Press, New York and London.
122. Owen, E. C. (1959). *Rev. Pathol. Gén. Physiol. Clin.* **59**, 231–43; (1959). *C.A.* **53**, 11457.
123. Beck, W. S. (1975). *Biochem. Pathophysiol.* 403–50.
124. Comar, C. L. and Bronner, F., Eds (1962). "Mineral Metabolism: An Advanced Treatise". Academic Press, New York and London.
125. Owen, E. C. (1960). *Proc. Nutr. Soc.* **19**, 154–62.
126. Beskid, M. (1962). *Polski Tygod. Lekar.* **17**, 525–9; (1965). *C.A.* **62**, 3143.
127. Bittel, R. (1968). Commis. Energ. At. (Fr.), Serv. Doc., Ser. "Bibliog"., CEA-BIB-130; (1969). *C.A.* **70**, 56713.
128. Reinhold, J. G. (1975). *Clin. Chem.* **21**, 476–500.
129. Underwood, E. J. (1971). "Trace Elements in Human and Animal Nutrition", 3rd Edn. Academic Press, New York and London.
130. Beeson, K. C. and Matrone, G. (1976). "The Soil Factor in Nutrition: Animal and Human", Nutrition and Clinical Nutrition. Marcel Dekker, New York.

10 Determination of Low Concentrations of Cobalt

In soils, fertilizers, waters, plant materials, and all parts of both the animal and human body, cobalt will almost invariably occur only in low concentrations. In industrial chemical analysis, cobalt is usually determined by procedures which give an accuracy of about 0.01%, or 100 parts 10^{-6}; in biological materials, the cobalt content is often only 1 part 10^{-6}, or even less. The simple expedient of increasing the sample size for a low concentration of an element is often impossible for biological substances. Furthermore, the frequent occurrence of a large quantity of organic matter in the sample requires that additional steps be taken in the decomposition stage. It is hoped that a brief outline of the principal procedures for determining a low content of cobalt, together with a selected list of recent references, will prove helpful to readers.

Sample Preparation

Soils and fertilizers, after drying, can be pulverized in conventional equipment to a fineness of -100 mesh, thoroughly mixed, and placed in clean glass or plastic sample containers.

Plant material may often require washing to remove adhering soil, but this operation must not be prolonged lest leaching occur. After drying at 105 °C, the sample can be ground to a suitable size in a hammer or ball mill designed for pulverizing plant products and similar organic material.

Most equipment in a sampling room, such as pulverizers and screens, will be virtually free of cobalt; contamination from this source can usually be ignored. Stainless steel generally contains traces of cobalt, and it is preferable to avoid screens made of this alloy. Tungsten carbide mortars and pestles must not be used when low cobalt contents are to be determined, because cobalt is the binder for tungsten carbide.

Water and other liquid samples are usually brought to the laboratory in plastic bottles. However, many metal ions, including cobalt, tend to become partially adsorbed on a plastic surface in a neutral or alkaline medium. The addition of 5 ml of nitric acid to a liter of sample will prevent this, but it is advisable to minimize the time between collection of a liquid sample and determination for traces of cobalt.

Sample Decomposition

As a general rule, the determination of low cobalt concentrations will require the complete solution of the sample by treatment with a mixture of hydrochloric, nitric, and sulphuric acids in a beaker, followed by a fusion of any residue with sodium carbonate in a platinum crucible, and its addition to the filtrate resulting from the acid attack. Another procedure involves an initial treatment with hydrofluoric and sulphuric acids, in a platinum dish, followed by a transfer to a beaker and further attack with hydrochloric and nitric acids; if a residue persists it must be fused with sodium carbonate. Other variations include fusion with potassium bisulphate or sodium peroxide.

Many samples in biological chemistry contain a large quantity of organic matter, which must be completely destroyed before chemical analysis can be undertaken. This can be effected by repeated evaporation on a hot plate with either a mixture of nitric and sulphuric acids, or aqua regia and sulphuric acid. The decomposition of organic matter by the mixture of nitric and sulphuric acids is usually a rather slow process, and it can be hastened by the addition, after attack by nitric and sulphuric has proceeded for some time, of perchloric acid. When hot and concentrated, the latter is a powerful oxidizer; it must never be used for the initial attack on organic matter because of its violently explosive tendencies in such conditions.

Destruction of organic matter can also be carried out by combustion in a muffle, leaving cobalt and other inorganic constituents as a small ash residue, ready for acid treatment or fusion. If the cobalt concentration dictates the use of a large sample, decomposition by ashing is easier and faster than by acid treatment. With the latter, large samples often froth badly, and the complete oxidation of carbonaceous material on the walls of the beaker is frequently a lengthy task. Cobalt-containing samples should be put in new, or unetched, dishes of either porcelain or silica, or in a platinum dish, placed in a cold muffle furnace and gradually heated to 500 °C. Slow heating is essential to avoid any mechanical loss which would result from rapid ignition. Organic matter is usually completely destroyed after 1 h at 500 °C; if, after cooling and stirring the ash with a platinum wire or glass rod, black particles are still visible, the sample must be returned to the muffle for another hour. The small residue of ash is finally brought completely into solution by acid treatment or fusion. It has been reported[1] that no significant loss of cobalt occurs by volatilization when dry ashing up to 1000 °C, but the writer recommends a temperature of 500 °C to avoid the possibility of loss from reaction with, or adsorption on, container surfaces.

Separations

After the sample has been completely decomposed, it is usually necessary to separate the cobalt from a few, or possibly many, other constituents before proceeding with its determination. There are many separation procedures;[2-5] the choice depends on both the type of sample and the analytical techniques employed.

The traditional method of separating cobalt from cations of Groups 1 and 2 by precipitating the latter two with hydrogen sulphide from a 5–10% by vol. solution of either hydrochloric or sulphuric acid, still occupies a prominent place in analysis.

One important separation method utilizes a zinc oxide precipitation, which yields cobalt, with nickel and manganese, in the filtrate, and most metals of Group 3 in the precipitate; subsequent precipitation of cobalt by 1-nitroso-2-naphthol is specific.[6-9]

Cobalt can be separated from a number of elements by the addition of potassium nitrite to a dilute acetic acid solution.[2,8,9]

Solvent extraction is extensively used in cobalt analyses. Cupferron-chloroform in cold dilute acid solution,[2,8,9] dithizone in carbon tetrachloride at pH 8–9,[8,10] 1-nitroso-2-naphthol or 2-nitroso-1-naphthol in chloroform,[2,7-9] ammonium thiocyanate in a mixture of amyl alcohol and ether,[2,8,9] and sodium diethyldithiocarbamate and chloroform in ammoniacal solution[8,9] are all examples of the methods used to isolate cobalt from a number of interfering elements. In cobalt analyses, an ether extraction of iron in 1:1 hydrochloric acid is frequently used to remove large quantities of this common element.[2]

Anion exchange resins, such as Dowex 1, 1X8, and 1X10, and cation exchange resins, such as Amberlite IRA 120, have been employed to isolate cobalt from iron, nickel, and other common interfering metals.[8,9] Small quantities of cobalt have been separated from other elements by paper chromatography.[7,8]

Determination of Cobalt

Most determinations of low concentrations of cobalt are carried out by the procedures of atomic absorption, colorimetry, and optical spectrography; other techniques used for the low ranges of this element are polarography, x-ray fluorescence, radiochemical methods, and catalytic techniques. These will be briefly discussed; the comprehensive list of recent references will provide details for those working in this field.

Atomic absorption spectrophotometry

In less than twenty years, atomic absorption spectrophotometry, or atomic absorption spectroscopy, has become one of the most popular analytical techniques. Theory and practice are fully described in many books.[11-22]

For cobalt, line 240·72 nm is used most frequently, but others such as 242·49, 252·14, 345·35, and 352·69 have been employed. Air–acetylene or nitrous oxide–acetylene can be used, and for many samples no interferences from common elements are encountered.

When the cobalt content is very low, it is desirable to separate the cobalt from the bulk of the accompanying elements. Such concentrating practices, e.g. that of solvent extraction, have been described in many papers.[23-31] Other publications have discussed various aspects of the determination of low cobalt contents in many materials.[10,32-44] In recent years, some analyses of cobalt by atomic absorption spectrophotometry have been carried out by replacing the gas flame by a heated graphite tube.[45-50] A modification of atomic absorption, termed atomic fluorescence spectrophotometry, has also been used to determine the lower ranges of cobalt.[19,22,23,51,52]

Colorimetry

Colorimetric procedures for small concentrations of many elements, including cobalt, have been used successfully for a long period. The number of papers published on colorimetric methods for cobalt is roughly equal to the combined total of papers on all other analytical procedures for this element. Unfortunately, most of these have been developed for a specific material, and interference by many common elements limits their general application. The colorimetric procedures using ammonium thiocyanate, nitroso-R-salt, and 2-nitroso-1-naphthol, have proved reliable over many years for a wide variety of materials; these methods have been described in detail in reference books and papers.[2,8,9,53,54]

One or more separations are nearly always required before the final colorimetric determination of cobalt. For a very low cobalt content, a collecting step using, for example, either 1-nitroso-2-naphthol or dithizone, may precede final measurement by nitroso-R-salt.[54-57]

Optical spectrography

Optical spectrography, or emission spectrography, is a valuable technique for measuring small quantities of cobalt. The principal cobalt lines used in quantitative spectrographic work are 228·62, 237·86, 304·40, 306·18, 333·34, 335·44, 345·35, 346·58, and 352·98 nm. The principles and applications of spectrochemical analysis are detailed in various reference books;[58-62] methods for cobalt are discussed in books and papers.[2,7-9,63]

Polarography

A polarographic procedure may be used for a low cobalt content, especially if the sample contains a large quantity of nickel, and only small amounts of most other metals. A separation or concentrating step is nearly always required before the final polarographic measurement. This may entail classical separations, solvent extraction, ion exchange, or other techniques.

The theory and applications of polarography are given in two monographs;[64,65] details for cobalt are found in reference books,[2,7-9] and papers.[66-71]

X-Ray fluorescence

X-Ray fluorescence, or x-ray spectrography, offers another technique for determining cobalt over a wide range. For low concentrations of cobalt, as with previous procedures, it is usually necessary to eliminate a number of other elements, either by one of the separating or concentrating methods, before the final measurement.

Principles and applications of this technique are discussed in reference books;[72-76] and papers have been published on the determination of low concentrations of cobalt in various materials.[77-81]

Radiochemical methods

Low cobalt contents can be determined by radiochemical methods. The theory and applications of this technique are covered in reference books,[82-84] and a number of papers on cobalt have appeared.[85-98]

Catalytic methods

Traces of cobalt have been determined by the catalytic effect of this element on certain chemical reactions, such as the oxidation of alizarin by hydrogen peroxide. This technique is sometimes termed the "kinetic" determination of cobalt. Earlier references to the method have been given in monographs on cobalt analysis;[7,8] additional information may be obtained from later papers.[99-113]

Miscellaneous procedures

Very small quantities of cobalt have been determined by electron probe analysis. The subject has been reviewed in an analytical treatise,[113] and the precision and detection limits for cobalt in sulphides are given in a journal publication.[115]

Occasionally, traces of cobalt have been determined by mass spectrography,[116] but the equipment and staff required limit the use of this technique to large research institutions.

Field Tests in Geochemical Prospecting

To aid the search for ore bodies, special procedures have been developed for rapid, approximate determinations under field conditions for low cobalt contents in waters, soils, sediments, and vegetation.[8,117,118]

Forms of Cobalt

It is sometimes necessary to differentiate cobalt in the metallic, sulphide, and oxide forms in soils, dusts, and biological residues. The procedures developed for this purpose in the metallurgical industry can be readily applied.[119]

Cobalt Standards

Unlike many other metals, cobalt is not available commercially in a "five-nines" grade. Most common cobalt salts are not quite suitable for primary standards, owing either to uncertainty over the state of hydration, or to the presence of significant quantities of impurities.

Cobalt in the metallic or powder form is readily obtainable at a purity of 99·8–99·88 %. If the highest reagent grade of either the chloride, nitrate, or sulphate of cobalt is used to prepare standards, the cobalt value should be verified by a careful electrolytic determination.

The well-known spectrographically standardized substances of Johnson, Matthey and Co. Ltd., of London, in which metallic impurities are reduced to below 0·001 %, include cobalt metal, and the nitrate, oxide, and sulphate. Provided their purity is checked, these materials, or those mentioned in the previous paragraph, will furnish satisfactory standards for all determinations of low concentrations of cobalt.

References

1. Raaphorst, J. G., van Weers, A. W., and Haremaker, H. M. (1974). *Analyst* **99**, 523–7.
2. Young, R. S. (1971). "Chemical Analysis in Extractive Metallurgy". Charles Griffin, London.
3. Dilts, R. V. (1974). "Analytical Chemistry: Methods of Separation". Van Nostrand, New York.
4. Miller, J. M. (1975). "Separation Methods in Chemical Analysis". Wiley-Interscience, New York.
5. Grushka, E., ed. (1976). "New Developments in Separation Methods". Dekker, New York.
6. Tombu, C. (1963). *Cobalt* **20**, 103–10; **21**, 185–9.
7. Pyatnitskii, I. V. (1965). "Analytical Chemistry of Cobalt". Nauka, Moscow.
8. Young, R. S. (1966). "The Analytical Chemistry of Cobalt". Pergamon, Oxford.
9. Young, R. S. (1970). *In* "Encyclopedia of Industrial Chemical Analysis", Vol. 10, 283–326, 348–73. Wiley-Interscience, New York.
10. Armannsson, H. (1977). *Analyt. Chim. Acta* **88**, 89–95.
11. Elwell, W. T. and Gidley, J. A. F. (1966). "Atomic-Absorption Spectrophotometry", 2nd Edn. Pergamon, Oxford.
12. Angino, E. E. and Billings, G. K. (1967). "Atomic Absorption Spectrometry in Geology". Elsevier, New York.
13. Slavin, W. (1968). "Atomic Absorption Spectroscopy". Interscience, New York.
14. Ramirez-Munoz, J. (1968). "Atomic-Absorption Spectroscopy and Analysis by Atomic-Absorption Flame Photometry". Elsevier, New York.
15. Rubeska, I. and Moldan, B. (1969). "Atomic Absorption Spectrophotometry". Butterworth, London.
16. Christian, G. D. and Feldman, F. J. (1970). "Atomic Absorption Spectroscopy: Applications in Agriculture, Biology, and Medicine". Wiley, Chichester.
17. L'vov, B. V. (1970). "Atomic-Absorption Spectrochemical Analysis". Adam Hilger, London.
18. Price, W. J. (1972). "Analytical Atomic Absorption Spectrometry". Heyden and Son, London.
19. Kirkbright, G. F. and Sargent, M. (1974). "Atomic Absorption and Fluorescence Spectroscopy". Academic Press, London and New York.
20. Robinson, J. W. (1975). "Atomic Absorption Spectroscopy". Edward Arnold, London.
21. Pinta, M., ed. (1975). "Atomic Absorption Spectroscopy". Halsted, New York.
22. Thompson, K. C. and Reynolds, R. J. (1978). "Atomic Absorption Fluorescence and Flame Emission Spectroscopy", 2nd Edn. Charles Griffin, High Wycombe.
23. Fleet, B., Liberty, K. V., and West, T. S. (1968). *Analyst* **93**, 701–8.
24. Steele, T. W., Wepener, H., and Taylor, J. D. (1968). Nat. Ins. Met. Rep. S. Africa, Res. Rep. No. 362.
25. Mountjoy, W. (1970). U.S. Geol. Survey Prof., Paper No. 700-B.
26. Donaldson, E. M. and Rolko, V. H. E. (1967). Dept. Energy, Mines and Resources, Ottawa, Tech. Bull. TB93.
27. Donaldson, E. M., Charette, D. J., and Rolko, V. H. E. (1969). *Talanta* **16**, 1305–10.
28. Jago, J., Wilson, P. E., and Lee, B. M. (1971). *Analyst* **96**, 349–53.
29. Simmons, W. J. (1973). *Analyt. Chem.* **45**, 1947–9.

30. Riley, J. P. and Taylor, D. (1968). *Analyt. Chim. Acta* **40**, 479–85.
31. Riley, J. P. and Topping, G. (1969). *Analyt. Chim. Acta* **44**, 234–6.
32. Scholes, P. H. (1968). *Analyst* **93**, 197–210.
33. Fleming, H. D. (1972). *Analyt. Chim. Acta* **59**, 197–208.
34. Cobb, W. D., Foster, W. W., and Harrison, T. S. (1972). *Analyt. Chim. Acta* **60**, 430–3.
35. Engberg, A. (1970). *Analyt. Chim. Acta* **50**, 531–2.
36. Fujiwara, K., Haraguchi, H., and Fuwa, K. (1972). *Analyt. Chem.* **44**, 1895–7.
37. Ure, A. M. and Mitchell, R. L. (1967). *Spectrochim. Acta* **23B**, 79–96.
38. Ward, F. N., Nakagawa, H. M., Harms, T. F., and Van Sickle, G. H. (1969). U.S. Geol. Survey Bull. 1289.
39. Sachdev, S. L., Robinson, J. W., and West, P. W. (1967). *Analyt. Chim. Acta* **38**, 499–506.
40. Kapetan, J. P. (1969). Am. Soc. Test. Mat. Spec. Tech. Publ. No. 443.
41. Levine, S. L. (1969). *At. Absorp. Newsl.* **8**, 58–9.
42. Christian, G. D. and Feldman, F. J. (1969). *Can. Spectrosc.* **14**, 80–3.
43. Blanco-Romia, M. and Gonzales Escoda, J. M. (1976). *Analysis* **4**, 177–9; (1976). *C.A.* **85**, 56224.
44. Chowdhury, A. A., De, A. K., and Das, A. K. (1977). *Z. Analyt. Chem.* **284**, 41.
45. Simmons, W. J. (1975). *Analyt. Chem.* **47**, 2015–18.
46. Muzzarelli, R. A. A. and Rochetti, R. (1975). *Talanta* **22**, 683–5.
47. Yan, J. (1975). *Ti Ch'iu Hua Hsueh* **4**, 291–6; (1976). *C.A.* **84**, 83746.
48. Merzlyakov, A. V., Malinina, R. D., Solomatin, V. T., and Nuzhdina, V. N. (1976). *Zavod. Lab.* **42**, 1331–2; (1977). *C.A.* **86**, 100433.
49. Nagdaev, V. K., Bukreev, Y. F., and Zolotavin, V. L. (1976). *Zavod. Lab.* **42**, 1333–5; (1977). *C.A.* **86**, 100434.
50. Matousek, J. and Sychra, V. (1969). *Analyt. Chem.* **41**, 518–22.
51. Norris, J. D. and West, T. S. (1971). *Analyt. Chim. Acta* **55**, 359–65.
52. Adam, J. and Pribil, R. (1971). *Talanta* **18**, 733–7.
53. Dewey, D. W. and Marston, H. R. (1971). *Analyt. Chim. Acta* **57**, 45–9.
54. Nazarenko, V. A. and Shitareva, G. G. (1958). *Zavod. Lab.* **24**, 932–4; (1960). *C.A.* **54**, 11836.
55. Beeson, K. C., Kubota, J., and Lazar, V. A. (1965). *Agronomy* **9**, 1064–77.
56. Forster, W. and Zeitlin, H. (1966). *Analyt. Chim. Acta* **34**, 211–24.
57. A.S.T.M. (1968). "Methods for Emission Spectrochemical Analysis: General Practices, Nomenclature, Tentative Methods, Suggested Methods". A.S.T.M., Philadelphia.
58. Blackburn, J. A. (1970). "Spectral Analysis: Methods and Techniques". Dekker, New York.
59. Grove, E. L. (1971). "Analytical Emission Spectroscopy", Part 1. Dekker, New York.
60. Straughan, B. P. and Walker, S. (1976). "Spectroscopy", Vols 1–3. Halsted, New York.
61. Barnes, R. M., ed. (1976). "Emission Spectroscopy". Wiley, Chichester.
62. Nichol, I. and Henderson-Hamilton, J. C. (1964–5). *Trans. Instn Min. Metall.* **74**, 955–61.
63. Heyrovsky, J. and Kuta, J. (1966). "Principles of Polarography". Academic Press, New York and London.
64. Crow, D. R. and Westwood, J. V. (1968). "Polarography". Barnes and Noble, New York.

65. Miklos, I. (1964). *Femip. Kut. Int. Közlen.* **7**, 437–40; (1966). *C.A.* **65**, 16054.
66. Temmerman, E. and Verbeck, F. (1970). *Analyt. Chim. Acta* **50**, 505–14.
67. Aleskovskii, V. B., Lapitskaya, S. K., and Sviridenko, V. G. (1972). *Zh. Analit. Khim.* **27**, 602–4; (1972). *C.A.* **77**, 28559.
68. Harvey, B. R. and Dutton, J. W. R. (1973). *Analyt. Chim. Acta* **67**, 377–85.
69. Bishara, S. W., Gawargious, Y. A., and Faltaoos, B. N. (1974). *Analyt. Chem.* **46**, 1103–5.
70. Vasil'eva, L. N. and Yustus, Z. L. (1976). *Zavod. Lab.* **42**, 911–12; (1977). *C.A.* **86**, 64990.
71. Blokhin, M. A. (1966). "Methods of X-Ray Spectroscopic Research". Pergamon, Oxford.
72. Birks, L. S. (1969). "X-Ray Spectrochemical Analysis", 2nd Edn. Wiley-Interscience, New York.
73. Bertin, E. P. (1970). "Principles and Practice of X-Ray Spectrometric Analysis". Plenum, New York.
74. Jenkins, R. and De Vries, J. L. (1970). "Practical X-Ray Spectrometry", 2nd Edn. Macmillan, London.
75. Liebhafsky, H. A. and Pfeiffer, H. G. (1972). "X-Rays, Electrons, and Analytical Chemistry". Wiley-Interscience, New York.
76. Hubbard, G. L. and Green, T. E. (1966). *Analyt. Chem.* **38**, 428–32.
77. Wybenga, F. T. (1965). *Appl. Spectrosc.* **19**, 193.
78. Lassner, E., Puschel, R., and Schedle, H. (1965). *Talanta* **12**, 871–81.
79. Cocozza, E. P. and Ferguson, A. (1967). *Appl. Spectrosc.* **21**, 286–90.
80. Gulacar, O. F. (1974). *Analyt. Chim. Acta* **73**, 255–64.
81. Rakovic, M. (1970). "Activation Analysis". Iliffe, London.
82. Kruger, P. (1971). "Principles of Activation Analysis". Wiley-Interscience, New York.
83. De Soete, D., Gijbels, R., and Hoste, J. (1972). "Neutron Activation Analysis". Wiley-Interscience, New York.
84. De Soete, D. and Hoste, J. (1964). Radiochem. Methods Anal., Proc. Symp., Salzburg, Vol. 2, 91–100; (1965). *C.A.* **83**, 12310.
85. Landgrebe, A. R., McClendon, L. T., and De Voe, J. R. (1964). Radiochem. Methods Anal., Proc. Symp., Salzburg, Vol. 2, 321–30; (1965). *C.A.* **63**, 12310.
86. Hilton, D. A. and Reed, D. (1965). *Analyst* **90**, 541–4.
87. Ikeda, N. and Muto, H. (1966). *Radioisotopes* **15**, 165–8.
88. Muzzarelli, R. A. A. (1966). *Talanta* **13**, 193–7.
89. Landgrebe, A. R., Gills, T. E., and De Voe, J. R. (1966). *Analyt. Chem.* **38**, 1265–6.
90. Motojima, K., Bando, S., and Tamura, N. (1967). *Talanta* **14**, 1179–83.
91. Robertson, D. E. (1968). *Analyt. Chem.* **40**, 1067–72.
92. Liessens, J. L., Dams, R., and Hoste, J. (1969). *Analyt. Chim. Acta* **45**, 313–18.
93. Claassen, H. C. (1970). *Analyt. Chim. Acta* **52**, 229–35.
94. Morse, R. S. and Welford, G. A. (1970). *Analyt. Chem.* **42**, 1100–2.
95. Csajka, M. (1971). Lucr. Conf. Nat. Chim. Anal. 3rd, Vol. 4, 23–8; (1972). *C.A.* **77**, 13525.
96. Benaben, P., Barrandon, J. N., and Debrun, J. L. (1975). *Analyt. Chim. Acta* **78**, 129–43.
97. Salbu, B., Steinnes, E., and Pappas, E. C. (1975). *Analyt. Chem.* **47**, 1011–16.
98. Schenk, G. H., Dilloway, K. P., and Coulter, J. S. (1969). *Analyt. Chem.* **41**, 510–14.

99. Gillard, R. D. and Spencer, A. J. (1970). *J. Chem. Soc. A* **10**, 1761–3.
100. Batley, G. E. (1971). *Talanta* **18**, 1225–32.
101. Costache, D. and Popa, G. (1971). *Anal. Univ. Buceresti Chim.* **20**, 83–8; (1973). *C.A.* **78**, 92169.
102. Vershinin, V. I., Chuiko, V. T., and Reznik, B. E. (1971). *Zh. Analit. Khim.* **26**, 1710–18; (1972). *C.A.* **76**, 41699.
103. Costache, D. (1971). *Rev. Roum. Chim.* **16**, 565–8; (1971). *C.A.* **75**, 44568.
104. Costache, D. (1971). *Rev. Roum. Chim.* **16**, 681–5; (1971). *C.A.* **75**, 58354.
105. Costache, D. (1972). *Farmacia* **20**, 545–51; (1973). *C.A.* **78**, 143425.
106. Reznik, B. E., Chuiko, V. T., and Vershinin, V. I. (1972). *Zh. Analit. Khim.* **27**, 395–7; (1972). *C.A.* **77**, 13594.
107. Janjic, T. J. and Milovanovic, G. A. (1972). *Glas. Hem. Drus., Beograd.* **37**, 173–89; (1973). *C.A.* **78**, 168149.
108. Trofimov, N. V., Kanaev, N. A., and Busev, A. I. (1974). *Zh. Analit. Khim.* **29**, 2001–17; (1975). *C.A.* **82**, 92646.
109. Ionescu, G. and Duca, A. (1974). *Bull. Inst. Polit. Iasi, Sect. 2* **20**, 15–24; (1975). *C.A.* **83**, 157387.
110. Pantaler, R. P., Alfimova, L. D., Bulgakova, A. M., and Pulyaeva, I. V. (1975). *Zh. Analit. Khim.* **30**, 946–50; (1975). *C.A.* **83**, 157326.
111. Pantaler, R. P., Alfimova, L. D., and Bulgakova, A. M. (1975). *Zh. Analit. Khim.* **30**, 1834–6; (1976). *C.A.* **84**, 69027.
112. Kreingol'd, S. U., Sosenkova, L. I., and Vzorova, I. F. (1976). *Metodȳ Anal. Kontrolya Proizvod. Khim. Prom-sti* **2**, 38–41; (1976). *C.A.* **85**, 136685.
113. Wittry, D. B. (1964). *In* "Treatise on Analytical Chemistry", Part 1, Vol. 5, 3173–232. Interscience, New York.
114. Heidel, R. H. (1972). *Analyt. Chem.* **44**, 1860–2.
115. Paulsen, P. J., Alvarez, R., and Mueller, C. W. (1970). *Analyt. Chem.* **42**, 673–5.
116. U.S. Geological Survey Bull. 1152, Washington, D.C. (1963).
117. Stanton, R. E. (1976). "Analytical Methods for Use in Geochemical Exploration". Edward Arnold, London.
118. Young, R. S. (1974). "Chemical Phase Analysis". Charles Griffin, London.

Subject Index